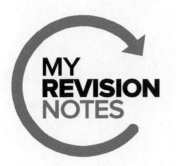

MY REVISION NOTES

AQA

A-level

GEOGRAPHY

SECOND EDITION

Helen Harris

Boost

HODDER
EDUCATION
AN HACHETTE UK COMPANY

Photo credits

p. 52 reused under a CC BY-SA 3.0 license; **p.74** © Michael Raw.

Although every effort has been made to ensure that website addresses are correct at time of going to press, Hodder Education cannot be held responsible for the content of any website mentioned in this book. It is sometimes possible to find a relocated web page by typing in the address of the home page for a website in the URL window of your browser.

Hachette UK's policy is to use papers that are natural, renewable and recyclable products and made from wood grown in well-managed forests and other controlled sources. The logging and manufacturing processes are expected to conform to the environmental regulations of the country of origin.

Orders: please contact Hachette UK Distribution, Hely Hutchinson Centre, Milton Road, Didcot, Oxfordshire, OX11 7HH. Telephone: +44 (0)1235 827827.
Email: education@hachette.co.uk. Lines are open from 9 a.m. to 5 p.m., Monday to Friday.
You can also order through our website: www.hoddereducation.co.uk

ISBN: 978 1 3983 2550 0

First edition published in 2017.
This edition published in 2021 by
Hodder Education,
An Hachette UK Company
Carmelite House
50 Victoria Embankment
London EC4Y 0DZ

www.hoddereducation.co.uk

Impression number   10 9 8 7 6 5 4 3 2

Year   2025 2024 2023

Cover photo © Cherries – stock.adobe.com

Illustrations by Aptara, Inc.

Typeset in India by Aptara, Inc.

Printed and bound by CPI Group (UK) Ltd, Croydon CR0 4YY

A catalogue record for this title is available from the British Library.

Get the most from this book

Everyone has to decide his or her own revision strategy, but it is essential to review your work, learn it and test your understanding. These Revision Notes will help you to do that in a planned way, topic by topic. Use this book as the cornerstone of your revision and don't hesitate to write in it — personalise your notes and check your progress by ticking off each section as you revise.

You can also keep track of your revision by ticking off each topic heading in the book. You may find it helpful to add your own notes as you work through each topic.

Tick to track your progress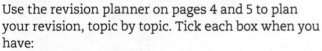

Use the revision planner on pages 4 and 5 to plan your revision, topic by topic. Tick each box when you have:
+ revised and understood the topic
+ tested yourself
+ practised the exam questions and gone online to check your answers and complete the quick quizzes.

Features to help you succeed

Exam tips

Expert tips are given throughout the book to help you polish your exam technique in order to maximise your chances in the exam.

Typical mistakes

The author identifies the typical mistakes candidates make and explains how you can avoid them.

Now test yourself

These short, knowledge-based questions provide the first step in testing your learning. Answers are at the back of the book.

Definitions and key words

Clear, concise definitions of essential key terms are provided where they first appear.

Key words from the specification are highlighted in bold throughout the book.

Making links

This feature identifies specific connections between topics and tells you how revising these will aid your exam answers.

Exam skills summary

These summaries highlight how specific skills identified or applicable in that chapter can be applied to your exam answers.

Revision activities

These activities will help you to engage with each topic in an interactive way.

Exam practice

Practice exam-style questions are provided for each topic. Use them to consolidate your revision and practise your exam skills.

Summaries

The summaries provide a quick-check bullet list for each topic.

Online

Go online to check your answers to the exam questions and try out the extra quick quizzes at **www.hoddereducation.co.uk/myrevisionnotesdownloads**

My Revision Notes: AQA A-level Geography Second Edition

My Revision Planner

Check your understanding and progress at **www.hoddereducation.co.uk/myrevisionnotesdownloads**

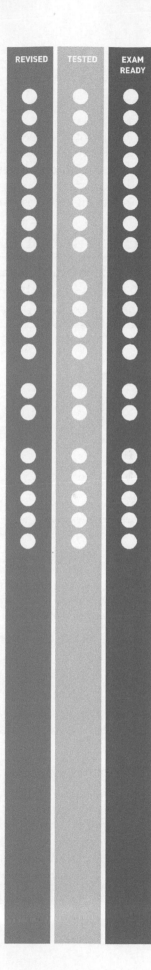

REVISED TESTED EXAM READY

My Revision Planner

My Revision Notes: AQA A-level Geography Second Edition

Introduction

As a student of A-level Geography it is important that you know about the:

+ structure of the exam
+ assessment objectives
+ key command words
+ key things to consider when preparing for and sitting the exam.

Structure of the exam

Paper 1. Physical Geography. Written examination. 2 hours and 30 minutes. 120 marks. 40% of the A-level

Section A	Section B	Section C
Question 1: Water and carbon cycles	**EITHER** Question 2: Hot desert systems and landscapes **OR** Question 3: Coastal systems and landscapes **OR** Question 4: Glacial systems and landscapes	**EITHER** Question 5: Hazards **OR** Question 6: Ecosystems under stress
Instructions		
Answer all questions. Total 36 marks.	Answer **EITHER** question 2, 3 **OR** 4. Total 36 marks.	Answer **EITHER** question 5 **OR** 6. Total 48 marks.
Question mark allocation		
One question worth 4 marks. Two questions worth 6 marks each. One question worth 20 marks.	One question worth 4 marks. Two questions worth 6 marks each. One question worth 20 marks.	Four multiple-choice questions worth 1 mark each. One question worth 6 marks. Two questions worth 9 marks each. One question worth 20 marks.
Types of question		

<td colspan="3">+ Short-answer questions worth 4 marks will test your ability to show knowledge and understanding. An example of a command word used would be **explain** or **assess**.
+ Questions worth 6 and 9 marks look for **knowledge** and **understanding** but also test your **application** of knowledge.
   + You could be asked to **analyse** and **interpret** patterns shown through various forms of data presentation, e.g. graphs, charts, photographs, written text and maps.
   + You may need to apply your knowledge to unfamiliar locations.
+ Questions worth 20 marks involve **extended prose**.
   + These are essay-style questions.</td>		

Timings		
Approximately 45 minutes	Approximately 45 minutes	Approximately 55 minutes

Paper 2. Human Geography. Written examination. 2 hours 30 minutes. 120 marks. 40% of the A-level

Section A	Section B	Section C
Question 1: Global systems and global governance	Question 2: Changing places	**EITHER** Question 3: Contemporary urban environments **OR** Question 4: Population and the environment **OR** Question 5: Resource security
Instructions		
Answer all questions. Total 36 marks.	Answer all questions. Total 36 marks.	Answer **EITHER** question 3, 4 **OR** 5. Total 48 marks.
Question mark allocation		
One question worth 4 marks. Two questions worth 6 marks each. One question worth 20 marks.	One question worth 4 marks. Two questions worth 6 marks each. One question worth 20 marks.	Four multiple-choice questions worth 1 mark each. One question worth 6 marks. Two questions worth 9 marks each. One question worth 20 marks.
Type of question		
<ul><li>Short-answer question worth 4 marks will test your ability to show knowledge and understanding. An example of a command word used would be **explain** or **assess**.</li><li>Questions worth 6 and 9 marks look for **knowledge** and **understanding** but will also test your **application** of knowledge.<ul><li>You could be asked to **analyse** and **interpret** patterns shown through various forms of data presentation, e.g. graphs, charts, photographs, written text and maps.</li><li>You may need to apply your knowledge to unfamiliar locations.</li></ul></li><li>Questions worth 20 marks involve **extended prose**.<ul><li>These are essay-style questions.</li></ul></li></ul>		
Timings		
Approximately 45 minutes	Approximately 45 minutes	Approximately 55 minutes

NB: The fieldwork investigation accounts for the remaining 20% of the A-level.

Assessment objectives

Assessment objectives	Requirements
AO1: Demonstrate knowledge and understanding	Show knowledge and understanding of places, environments, concepts and geographical processes at a variety of scales from local to global.
AO2: Apply knowledge and understanding	Demonstrate the ability to interpret, analyse and evaluate geographical information and issues.
AO3: Use a variety of relevant quantitative, qualitative and fieldwork skills.	Show the ability to use a range of key skills to interpret, analyse and evaluate geographical data and evidence.

Command words

+ Evaluate – often used in questions requiring extended writing. It requires you to examine an issue from different points of view with a focus on the strengths and weaknesses.
+ Analyse – to provide an in-depth account of something.

7

+ Assess – to consider several options or points of view and weigh them up so as to come to a conclusion regarding their effectiveness.
+ To what extent – often used for questions requiring extended writing. It means that you need to form an opinion on the merit or validity of a statement used in the question. You need to present evidence (facts and knowledge) and address the different sides of a statement or argument.
+ Discuss – set out both sides of an argument (for and against) and form a conclusion related to the content of your answer. Your answers should balanced.

Things to consider when preparing for and sitting the exam

Timing

+ Timing is a key factor for Papers 1 and 2.
+ The timings given are approximate. 45 minutes is just under 1 mark per minute and equates to spending about 10 minutes on a 9-mark question.

Data analysis and response

+ For several questions you will have to analyse data as part of your response so you must allow time to read and interpret the data resource – **but** be careful not to spend too long on this. Some of the resources are quite complex and it is easy to spend a long time trying to interpret the information and not leave enough time to compose your answer.
+ Make some annotations.
+ Make sure that you have practised plenty of data response questions; this is an essential examination skill.
+ Data and resources provided in the exam will be location specific, but you do not need to have studied the particular location. What you need is to be able to apply your knowledge to these unfamiliar locations (a synoptic link).
+ Never just repeat data that is in front of you. You need to:
    + add some narrative to explain what the data shows
    + manipulate the data
    + draw out general trends
    + identify anomalies
    + look for links across data sets where appropriate.

Essay questions

+ Essay questions will require you to discuss concepts with command words such as 'to what extent ...' or 'evaluate the success of ...', or 'do you agree with the view that ...?'.
+ Essays require some brief planning. The question will have several components and these need to be broken down when planning your response.
+ In extended answers, always refer to case studies and examples where possible.
+ Present different points of view in your essays.

Case studies

+ Case studies could be examined through any question.
+ Place-specific facts are key to good use of case study material.
+ With all topics, make sure that your facts are up to date.

Check your understanding and progress at **www.hoddereducation.co.uk/myrevisionnotesdownloads**

1 Water and carbon cycles

Water and carbon cycles as natural systems

REVISED ●

Systems in physical geography

In physical geography, two general approaches are used for explanation: models and systems.

+ A **model** is an idealised representation of reality.
+ A **system** is a set of interrelated events or components working together. It consists of:
  + inputs
  + stores
  + outputs
  + a series of flows or connections between the inputs, stores and outputs.

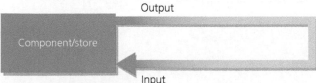

Figure 1.1 A closed system

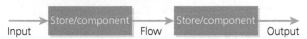

Figure 1.2 An open system

Systems can be classified as:

+ **isolated:** there are no interactions with anything outside the system boundary – there is no input or output of energy or matter
+ **closed:** there is transfer of energy into and beyond the system but no transfer of matter (see Figure 1.1)
+ **open:** both energy and matter transfer freely into and out of the system (see Figure 1.2)
+ **subsystem:** a component of a larger system. The Earth system has four subsystems, each of which is an open system with interrelationships between them (see Figure 1.3).

> **Input** The addition of matter and/or energy into a system.
>
> **Store** A part of the system where energy/mass is stored or transformed.

> **Exam tip**
>
> Systems are a core concept in physical geography. They must be understood at a variety of **scales**. For example, for water:
> + global hydrological cycle
> + drainage basin system
> + hill slope drainage system.

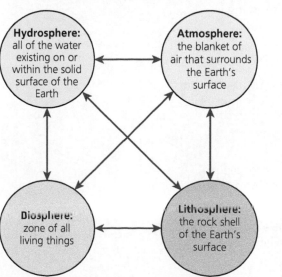

Figure 1.3 The four subsystems of the Earth system

Dynamic equilibrium is where there is a balance between inputs and outputs. For example, wave currents remove and replace sand on a shoreline but the beach apparently stays the same.

Feedback occurs when a change in one part of the system causes a change in another part. There are two types of feedback:

+ **Negative feedback:** a feedback which keeps a system in its original condition – for example, increase in CO_2 → increase in temperature → increased plant growth → increased uptake of CO_2 → reduction in CO_2, which counterbalances the initial increase.
+ **Positive feedback:** a feedback where there is a progressively greater change from the original condition of the system – for example, increase in temperature → increase in oceanic temperature → dissolved CO_2 released from warmer oceans → increase in CO_2 → further atmospheric warming.

Exam tip

The concepts of positive and negative feedback must be applied to a range of concepts. Correct sequencing is important – remember, the same catalyst can lead to both positive and negative feedbacks.

Now test yourself

TESTED

1 What is the difference between an open and a closed system?
2 Explain the links between the following subsystems:
   a) atmosphere and hydrosphere
   b) lithosphere and biosphere
   c) cryosphere and lithosphere.

Answers on p. 258

Application of the system concept to the water and carbon cycle

Four vital cycles connect the Earth's subsystems. These are the:

+ water cycle
+ carbon cycle
+ oxygen cycle
+ nitrogen cycle.

They are all fundamental to life on Earth and to a study of physical geography. Both the carbon and the water cycles are under pressure from growing populations and climate change.

Making links

Atmospheric CO_2 levels have a direct link to air temperature and thereby all major water stores.

Revision activity

Create flow diagrams to explain how different catalysts (e.g. increased water vapour) lead to positive and/or negative feedback in the water and carbon cycles. An example for the water cycle is:

Temperature increase → more evaporation → more water vapour in the atmosphere → greater cloud cover and more → greater absorption of long-wave radiation → further temperature increase. Positive feedback.

Check your understanding and progress at **www.hoddereducation.co.uk/myrevisionnotesdownloads**

The water cycle

Global distribution and size of major water stores

About 71 per cent of the Earth's surface is covered in water. The sizes of the world's water stores are shown in Figure 1.4.

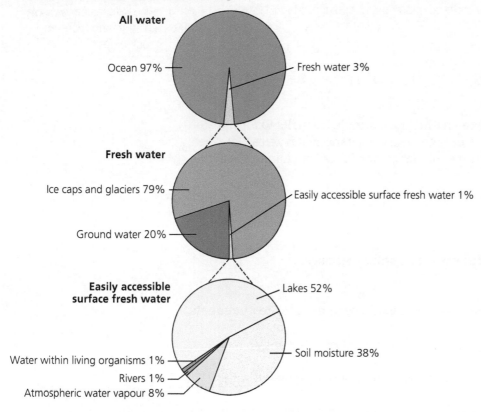

Figure 1.4 Sizes of major water stores

Now test yourself

3 Explain how growing populations are increasing pressure on ground water stores.

Answer on p. 258

TESTED ○

Exam tip

It is important to be able to recognise and explain the pressure from growing populations and climate change on water stores in:
+ the ocean
+ fresh water
+ soil moisture.

Water exists in three states:
+ solid (ice)
+ liquid
+ gas (water vapour).

The types of water and their global distribution are outlined in Table 1.1.

Table 1.1 Types of water and global distribution

Type of water store	Description
Oceanic water	+ There are five oceanic bodies of water and several smaller seas covering approximately 71% of the Earth's surface. + The Pacific Ocean is the largest.
Cryospheric water	+ Composed of sea ice, ice caps, ice sheets, Alpine glaciers and permafrost. + Mainly in high-altitude and high-latitude areas, including ice sheets of Antarctica, Greenland, Arctic areas of Canada and Alaska; ice caps such as the Himalayas, the Rockies, the Andes and the southern Alps of New Zealand.
Terrestrial water	+ Rivers, the largest by discharge of water being the Amazon. Lakes – Canada and Finland have the largest number of lakes. + Wetlands, where water covers the soil – these are present on every continent except Antarctica. + Groundwater, soil water and biological water also make up terrestrial water.
Atmospheric water	+ The most common form is water vapour. + Important as it absorbs and reflects incoming solar radiation. + Warm air holds more water vapour than cold air – a small increase in water vapour will lead to an increase in atmospheric temperatures (positive feedback, see page 30).

Processes driving change in water stores over time and space

Water changes from one state to another, for example ice melts to form water (latent heat is needed), water freezes to form ice (latent heat is released). The following processes are key to an explanation of how water changes from one state to another:

+ evaporation
+ condensation
+ cloud formation
+ precipitation
+ cryospheric processes (at different sizes and timescales).

Evaporation

+ **Evaporation** is a physical process where liquid becomes gas. Requirements include:
    + heat energy, provided either by the movement of water or by solar energy
    + air that is not saturated and can therefore absorb evaporated water molecules/water vapour.
+ **Transpiration** is linked to evaporation. It is a biological process where water is lost from plants through pores called stomata.
+ Together, the two processes are termed **evapotranspiration**. Factors affecting these processes include:
    + temperature
    + wind
    + humidity
    + climatic factors such as hours of sunshine.

Condensation

+ **Condensation** is a physical process where gas (water vapour) becomes liquid.
+ It happens when air cools and is less able to hold water vapour (dew point).
+ In the cooling process the water molecules condense onto nuclei (dust, smoke) or onto surfaces, for example grass, and form water droplets or frost.
+ **Precipitation** (rain, sleet, snow, hail) occurs when the air can no longer hold the weight of the condensed water.

Cloud formation

Clouds are visible masses of water droplets or ice crystals held in the atmosphere. They form when:

+ air is saturated either because it has cooled below the dew point or evaporation means the air has reached its maximum water-holding capacity
+ condensation nuclei are present.

The greater the amount of moisture in the cooling air, the greater the condensation and cloud formation.

Precipitation

The condensation which is a direct cause of precipitation can occur when:

+ air temperature is reduced to dew point, for example warm moist air passes over a cold surface on a clear night, or when heat is radiated out into the atmosphere and the ground gets colder, cooling the air above it
+ volume of air increases as it rises and expands, but there is no addition in heat (adiabatic cooling). In this example, the air may be forced to rise for three different reasons, each resulting in precipitation:
    + Air is forced to rise over hills and mountains = **orographic rainfall**.
    + Air masses of different temperatures and densities meet, the warm air rising over the cool sinking air = **frontal rainfall**.
    + Warm air rises from hot surfaces on a sunny day = **convectional rainfall**.

Now test yourself   TESTED ◯

8 Explain how the processes of evaporation and condensation relate to the formation of:
   a) clouds
   b) rainfall
9 How do the following affect evapotranspiration?
   a) temperature
   b) wind
   c) humidity
10 Why do sunny days lead to bursts of heavy rainfall?
11 Explain the link between climate change and soil moisture stores.

Answers on p. 258

Revision activity

Create a diagram to summarise the processes involved when water changes state.

Cryospheric processes

These affect the mass of ice at any scale. They include:

+ **accumulation** – inputs to a glacial system due to snowfall
+ **ablation** – output of a glacial system due to melting
+ **sublimation** – ice changing directly into water vapour.

Drainage basins as open systems

A **drainage basin** is an area of land drained by a river and its tributaries. Its boundary, or **watershed**, is marked by ridges of high land, beyond which rainfall will drain into a neighbouring drainage basin.

The drainage basin system:

+ forms a subsystem of the hydrological or water cycle
+ is an open system as it has inputs and outputs of both matter and energy
+ is composed of:
    + inputs (precipitation)
    + flows/transfers (throughfall, stemflow, infiltration, percolation, overland flow and groundwater flow)
    + stores (vegetation store)
    + outputs to the sea or atmosphere (evapotranspiration) – see Figure 1.5.

> **Flow/transfer** A form of linkage between one store/component and another that involves movement of energy or mass.

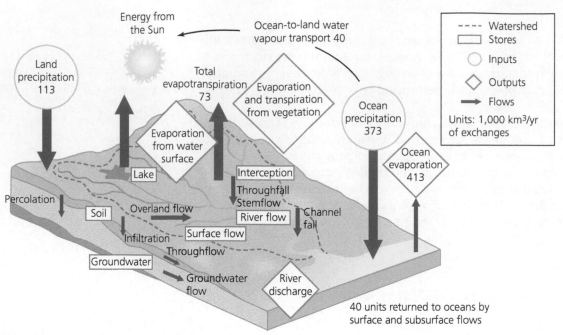

Figure 1.5 The drainage basin system

Table 1.2 Key terms for the drainage basin system

Precipitation	May fall as rain, hail, sleet or snow. Duration and intensity will impact processes in the system.
Evapotranspiration	Combined loss of water through evaporation and transpiration by plants.
Runoff	The output of water from the drainage basin system as it moves across the land surface either as overland flow or channel flow.
Interception store	Vegetation cover intercepts the precipitation and a store may be held on leaves and branches. Density of vegetation will affect this. Tropical rainforest can intercept 58% of rainfall.
Surface storage	This mainly occurs in built environments as puddles. In natural environments, infiltration normally occurs more quickly than rainfall and there will be surface puddles only after very long periods of rainfall or on compacted surfaces or bare rock.
Soil water storage	Pore spaces between soil particles fill with air and water. The amount of pore space varies in different soils: clay 40–60% volume, sand 20–45% volume.
Groundwater store	Water stored underground in permeable and porous rocks.
Channel store	The volume of water in a river channel.
Vegetation store	Vegetation cover intercepts the precipitation and a store may be held on leaves and branches. Density of vegetation will affect this. (Sometimes referred to as interception store.)
Stemflow	Water flows down the stems of plants and trees.
Infiltration	Water soaks into the soil. Rate = infiltration rate. The texture, structure and organic content of soil all affect infiltration rate. The rate normally declines during the early part of a storm.
Overland flow	Rainfall flowing over the ground surface either because the soil is saturated or because the rainfall is exceeding the soil infiltration capacity.
Channel flow	The flow of water in rivers.
Throughfall	Water moving from vegetation to the ground.
Throughflow	The lateral movement of water down a slope to a river channel. Slower than overland flow but the rate is decreased by root systems of vegetation.
Percolation	Downward movement of water into underground stores.
Groundwater flow	Downward and lateral movement of water within saturated rock. This is a very slow movement. Water-bearing rocks are called aquifers.
Evaporation	The process by which water, which is a liquid, changes to a gas. This process requires energy in the form of sun, aided by the wind.
Transpiration	The loss of water from vegetation through pores (stomata) on the surfaces.
River discharge	Volume of water in a river flowing past a certain point every second. Expressed in cumecs ($m^3 S^{-1}$).

Inputs   Flows   Stores   Outputs

Check your understanding and progress at **www.hoddereducation.co.uk/myrevisionnotesdownloads**

Now test yourself

TESTED ⬤

12 State three factors that can affect stores in the drainage basin system.

13 What leads to an increase in overland flow?

14 Explain the impact of increased ocean evaporation on the hydrological cycle.

Answers on p. 258

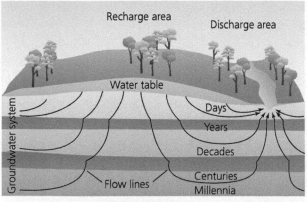

Figure 1.6 Varying timescales in the drainage basin system

Now test yourself

TESTED ⬤

15 Figure 1.6 shows that drainage basin systems change over time. Explain the seasonal impacts on the drainage basin system.

Answer on p. 258

Water balance

The long-term balance between the inputs and outputs in the drainage basin system is known as the **water balance**. It is expressed in an equation as:

$$P = Q + E \pm (S)$$

P = precipitation

Q = runoff (measured in river discharge)

E = evapotranspiration

S = change in storage

+ **Positive water balance** – precipitation exceeds evapotranspiration.
+ **Negative water balance** – evapotranspiration exceeds precipitation.

Storage affects water balance, for example in:

+ winter when precipitation is likely to be high, the soil storage may lead to a soil moisture **surplus** and increased runoff – positive water balance
+ summer when by utilisation of water by humans and vegetation is likely to be high, there may be a soil moisture **deficit** – negative water balance
+ autumn when initial precipitation will recharge the soil store.

Figure 1.7 applies these terms to a soil water budget graph.

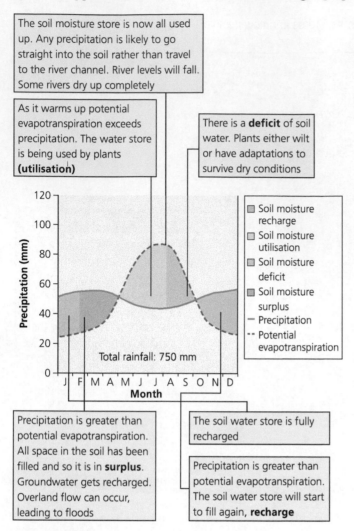

The soil moisture store is now all used up. Any precipitation is likely to go straight into the soil rather than travel to the river channel. River levels will fall. Some rivers dry up completely

As it warms up potential evapotranspiration exceeds precipitation. The water store is being used by plants (utilisation)

There is a **deficit** of soil water. Plants either wilt or have adaptations to survive dry conditions

Precipitation is greater than potential evapotranspiration. All space in the soil has been filled and so it is in **surplus**. Groundwater gets recharged. Overland flow can occur, leading to floods

The soil water store is fully recharged

Precipitation is greater than potential evapotranspiration. The soil water store will start to fill again, **recharge**

Legend:
- Soil moisture recharge
- Soil moisture utilisation
- Soil moisture deficit
- Soil moisture surplus
- Precipitation
- Potential evapotranspiration

Total rainfall: 750 mm

Figure 1.7 Annotated soil water budget graph

Now test yourself   TESTED ⃝

16 Explain the term 'water balance'.

Answer on p. 258

Answer on p. 258

Runoff variation

The flows of water within a drainage basin system end up either in the river (which then transfers the water by channel flow) or in groundwater stores. The water leaving the drainage basin through channel flow is called **runoff**.

Drainage basins all have individual characteristics. Interpreting the runoff variation and seasonal changes of a river's flow in a particular drainage basin can provide vital information for the management of water resources.

River discharge

River discharge is the measure of river flow (the volume of water passing a measuring point, measured in cumecs and calculated by multiplying cross-sectional area by velocity).

The river regime

The **river regime** is the annual variations in the amount of discharge in a river in response to climatic factors and drainage basin characteristics. It can be plotted on a **hydrograph** (a graph showing river discharge against time (see Figure 1.8)).

Exam tip

When presented with data to be used in an answer, it is essential that you 'add something new' and do not simply present information which is already given in the resource. You must add your analysis and interpretation of the facts, as shown by the annotations in Figure 1.7.

Revision activity

Using Figure 1.7, annotate a simple diagram version of the soil water budget graph with factors that will affect soil water budget (e.g. the effect of a summer drought or a prolonged winter freeze).

Now test yourself

17 List the factors that lead to runoff variations between different river regimes.

Answer on p. 258

Flood hydrograph

A flood hydrograph is a particular type of hydrograph which plots changes in the discharge of a river in response to a storm or rainfall event. It therefore represents a short-term event, not a long-term change. The key features of a flood hydrograph are shown in Figure 1.8.

Flood hydrograph A graph showing river discharge over a period of time when the river's normal flow is affected by a flood event.

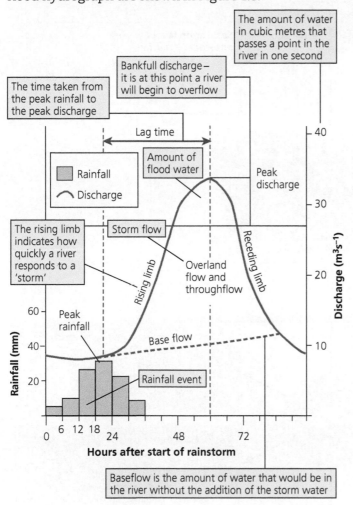

Figure 1.8 A flood hydrograph

Flood hydrographs can be described as:
+ **flashy:** short lag time, high peak, steep rising and falling limbs
+ **subdued:** long lag time, low peak, gently rising and falling limbs.

A range of physical and human factors will affect the response of a river to a storm event. They are summarised in Figure 1.9.

Lag time The time between peak rainfall and peak discharge.

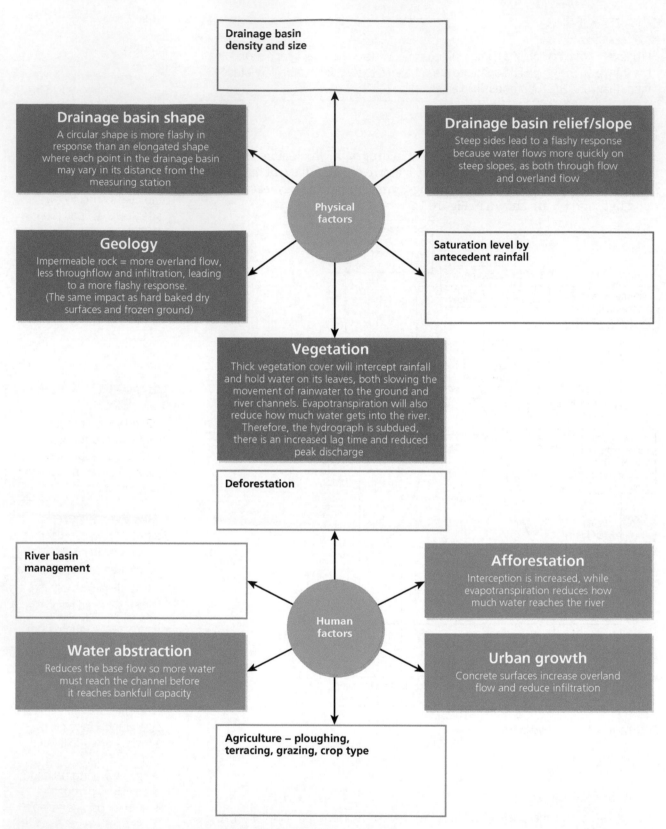

Figure 1.9 The physical and human factors affecting a flood hydrograph

Changes in the water cycle over time

The water cycle can be affected by both short- and long-term changes.

Short-term changes include:

+ daily fluctuations in temperature and rainfall – processes such as evaporation and transpiration are lower in cooler temperatures early in the morning and evening
+ seasonal changes – solar radiation varies between seasons and this impacts evapotranspiration and precipitation, which in turn impact the water cycle. Convectional heating in the summer months can cause heavy downpours.

Revision activity

Complete the blank boxes in Figure 1.9 to explain how each factor will affect a river's response to a storm.

Check your understanding and progress at **www.hoddereducation.co.uk/myrevisionnotesdownloads**

Long-term changes include the impacts of climate change and weather patterns over a number of years. For example, glacial periods lead to more surface ice storage and lower sea levels, slowing down the water cycle.

Table 1.3 outlines the effects of both human impact and natural variation on the water cycle.

Table 1.3 The effect of natural variation and human impact on the water cycle

Storm events	Seasonal changes

✦ Storm events can include flash floods and unseasonal or unexpected weather events.
✦ As air temperatures rise there is an increase in evaporation and an increase in the amount of water vapour held in the atmosphere. This leads to intense rainstorms where, due to the nature of the rainfall, there is:
  ✦ less infiltration
  ✦ more surface runoff
  ✦ more flooding.
✦ When water vapour condenses into rainfall it releases heat energy and this drives a stronger intensity of storm.
	✦ In wet seasons precipitation exceeds evapotranspiration, which leads to a water surplus. Groundwater stores are full and more surface runoff results in higher discharge levels in rivers.
	✦ In dry seasons precipitation is lower than evapotranspiration and groundwater stores are depleted, water used by humans and plants is not replaced, so there is a water deficit.

Farming practices – soil drainage	**Water abstraction, for example the London Basin**

✦ Farmers use a system of corrugated plastic tubing to drain water from soils when the water table in the soil is high. Improving drainage in poorly drained soils can increase productivity.
✦ Soil drainage can:
  ✦ improve soil structure
  ✦ aerate soils
  ✦ allow microorganisms to thrive and produce more organic matter
  ✦ reduce compaction from heavy machinery.
✦ It can also artificially increase throughflow, leading to flooding, and the dry surface layer can become prone to wind erosion.
	✦ Groundwater abstraction is the process of taking water from a ground source either temporarily or permanently.
	✦ Most of this water is used for irrigation but also, after treatment, for drinking water.
	✦ Hydrogeology is used to monitor safe levels of water abstraction, as over-abstraction can lead to a number of issues, including:
	  ✦ rivers drying up
	  ✦ damage to wetland ecosystems
	  ✦ sinking water tables
	  ✦ empty wells.
	✦ In coastal areas, intrusion of salt water from the sea degrades groundwater and leads to difficulties of usage for domestic and agricultural purposes.

Localised deforestation	**Extensive deforestation, for example the Amazon Basin**
✦ Evapotranspiration is lower as reduced vegetation cover has fewer leaves and fewer roots. ✦ There is less interception due to reduced canopy. ✦ Overland flow and throughflow increase as there is a lack of vegetation to slow down these processes. ✦ There is increased river discharge and risk of localised flooding.	✦ Most of the water leaves the area in channel flow rather than being recycled to the atmosphere by the process of evapotranspiration. ✦ A reduction of water vapour in the atmosphere leads to falling levels of precipitation. ✦ River levels fall.
Urbanisation	**Afforestation**
✦ Natural surfaces are replaced by concrete, so there is less interception and more overland flow. ✦ Infiltration is reduced so subsurface stores and flows are reduced. ✦ Water levels in rivers rise rapidly due to more overland flow.	✦ Plantations of natural forest increase interception and interception storage. ✦ Moisture held as interception storage evaporates back into the atmosphere. ✦ Runoff is reduced.

Revision activity

This is a peer group revision activity. Divide the factors affecting the water cycle in Table 1.3 among a small group, for instance three people each take two factors. Produce a flow diagram that explains the cause and effect of each impact and explain this to the rest of the group.

Consider exam practice question number 5, page 34, and as a group prepare a suitable response.

The water cycle and scale

The processes driving change in the water cycle vary over time and scale.

Global scale

+ Water is present on Earth as liquid, ice or atmospheric moisture.
+ It is cycled between these stores by a range of processes outlined above and summarised in the global water cycle shown in Figure 1.10.

Revision activity

Identify and make a list of the links between the water cycle and your chosen physical options topic, hazards or ecosystems under stress.

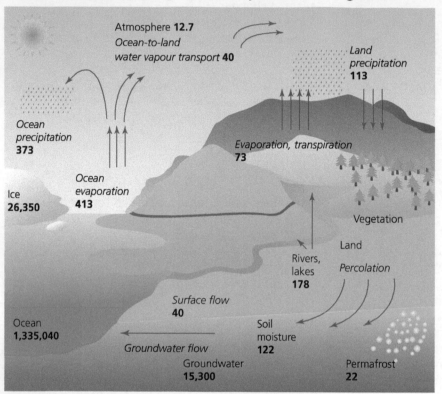

Figure 1.10 The global water cycle

Circulation of water between continents and oceans

+ Circulation of water is more rapid in tropical landmasses.
+ Most of the water in the Pacific Ocean recirculates within the Pacific itself.
+ Most of the water transported to the continents (North and South America, Europe and Africa) comes from the Atlantic Ocean.

Circulation of water in the drainage basin system

+ Each drainage basin will have its own unique characteristics depending on climate, geology, slope, soil, land use, vegetation, drainage basin shape and drainage density.
+ These characteristics determine the inputs, stores, flows and outputs at any individual location and the timescale of the processes involved – Figure 1.5.

Revision activity

Research a named example of a river in each of the following continents. You will need to think about climate, for example, and seasonal variations in temperature and rainfall:
+ Africa
+ South America
+ Asia.

For each named example, give a brief summary of the functions of its drainage basin system.

Now test yourself

20 Why is the circulation of water more rapid in tropical landmasses?

Answer on p. 258

TESTED

Check your understanding and progress at **www.hoddereducation.co.uk/myrevisionnotesdownloads**

The carbon cycle

Global distribution and size of major stores of carbon

Cycling of the element carbon is vitally linked to life on Earth. Carbon is present in molecules that are found in all living creatures. It is also present in the Earth's crust (sedimentary rocks, graphite, coal, oil and natural gas), atmosphere, soils and oceans.

When viewing the Earth as a system, these components can be referred to as stores of carbon (see Figure 1.11).

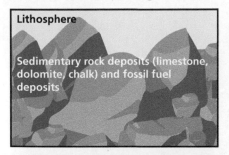

Lithosphere

Sedimentary rock deposits (limestone, dolomite, chalk) and fossil fuel deposits

Hydrosphere

Oceans:
- Surface layer – photosynthesis by plankton
- Intermediate and deep layer – carbon passes through the marine food chains and sinks to the ocean bed, where it is decomposed into sediments
- Living and dissolved organic matter
- Calcium carbonate shells in marine organisms

Terrestrial/biosphere
- Organic matter in soils, plant litter, soil humus and peat
- As organic molecules in living and dead organisms

Atmosphere

CO_2 gases in the atmosphere – a 'trace' gas accounting for 0.04% of the atmosphere, but this does not reflect its importance to life on Earth and the fact that CO_2 is a potent greenhouse gas that plays a vital role in regulating the Earth's surface temperature

Figure 1.11 Major stores of carbon

The sizes of global carbon stores are (approximately) as follows:
+ Atmosphere 600 Gt (gigatonnes, 1 Gt = 1 billion tonnes)
+ Ocean surface 700 Gt
+ Ocean deep layer 38,000 Gt
+ Sedimentary rocks 60,000,000–100,000,000 Gt
+ Soil 2,300 Gt
+ Terrestrial biomass 560 Gt
+ Fossil fuels 4,130 Gt

Now test yourself   TESTED

21 How is carbon stored in the biosphere and hydrosphere?

22 Where is the Earth's largest store of carbon?

Answers on p. 258

Processes driving change in the magnitude of carbon stores over time and space

The carbon cycle describes the transfer of carbon from one store/pool to another. At its simplest level it is expressed as shown in Figure 1.12.

In order to understand the cycling of carbon it is first important to understand not only where carbon is stored (**pools**) but also how long it stays there and the processes that transfer it from one pool to another (**fluxes**).

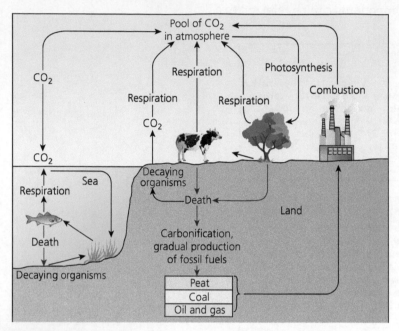

Figure 1.12 The carbon cycle

Figure 1.13 shows the stores and the fluxes (arrows). Purple represents pools/stores and red the processes that drive the transfer of carbon between the main pools.

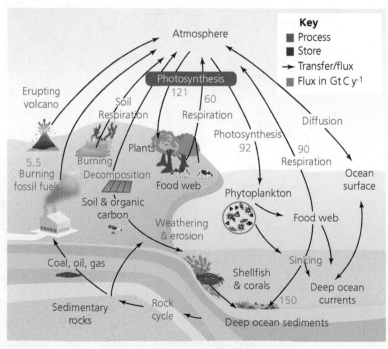

Figure 1.13 The carbon cycle: pools and fluxes

Because the quantities of carbon in the Earth's major pools are quite large, units such as petagrams, also known as a gigatonne (Gt), are used for the very large numbers involved.

Check your understanding and progress at **www.hoddereducation.co.uk/myrevisionnotesdownloads**

Now test yourself   TESTED ○

23 What are fluxes? Give three examples.

24 What is the name of the process by which CO_2 is transferred from the atmosphere to plants?

25 Combustion releases carbon into the atmosphere. What is the source of this carbon?

Answers on p. 258

Revision activity

Based on Figure 1.12, draw your own simplified diagram of the carbon cycle. Use colours to show:

+ processes
+ stores
+ transfers.

Subsystems of the carbon cycle

The subsystems of the carbon cycle are shown in Figure 1.14.

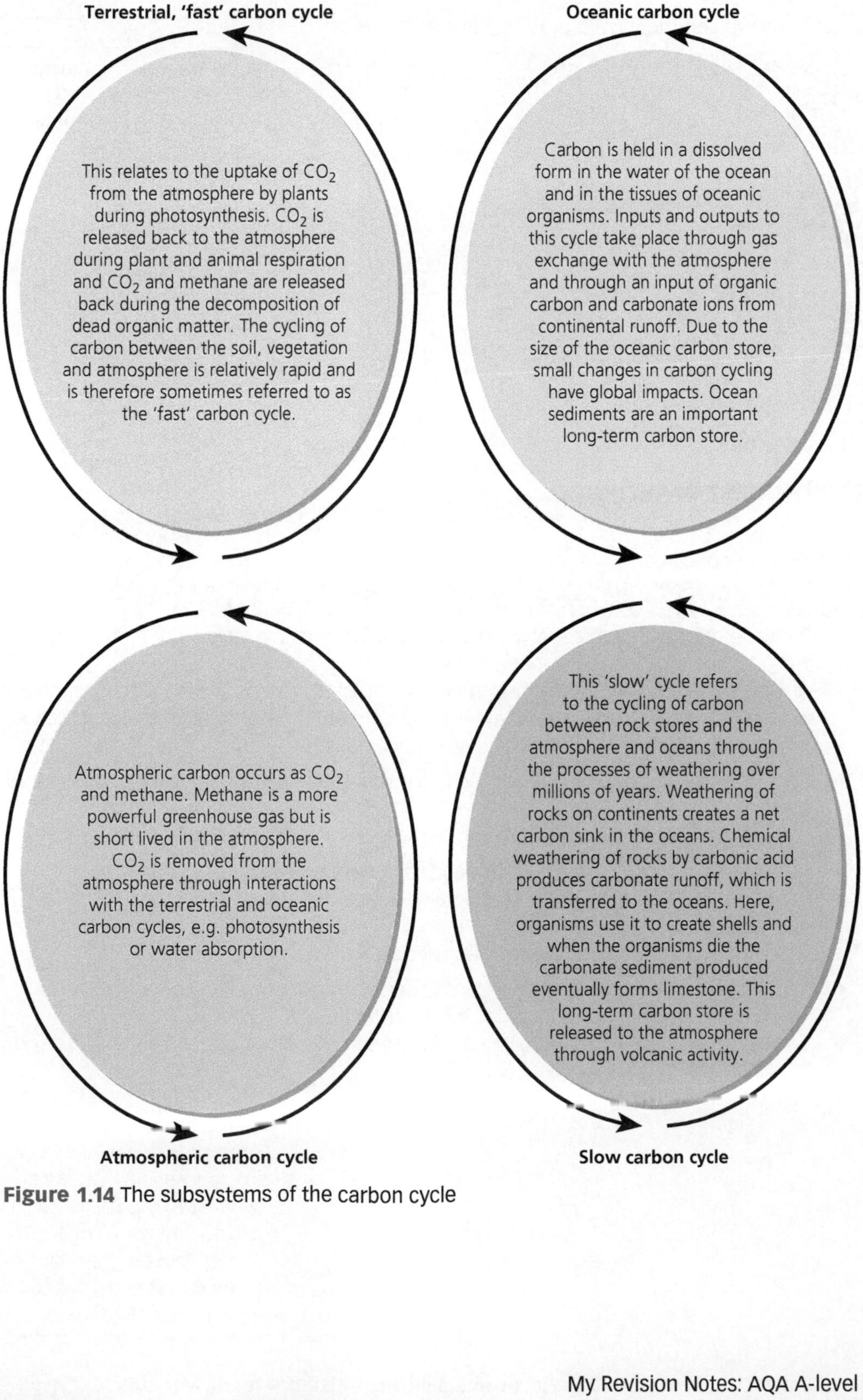

Terrestrial, 'fast' carbon cycle

This relates to the uptake of CO_2 from the atmosphere by plants during photosynthesis. CO_2 is released back to the atmosphere during plant and animal respiration and CO_2 and methane are released back during the decomposition of dead organic matter. The cycling of carbon between the soil, vegetation and atmosphere is relatively rapid and is therefore sometimes referred to as the 'fast' carbon cycle.

Oceanic carbon cycle

Carbon is held in a dissolved form in the water of the ocean and in the tissues of oceanic organisms. Inputs and outputs to this cycle take place through gas exchange with the atmosphere and through an input of organic carbon and carbonate ions from continental runoff. Due to the size of the oceanic carbon store, small changes in carbon cycling have global impacts. Ocean sediments are an important long-term carbon store.

Atmospheric carbon cycle

Atmospheric carbon occurs as CO_2 and methane. Methane is a more powerful greenhouse gas but is short lived in the atmosphere. CO_2 is removed from the atmosphere through interactions with the terrestrial and oceanic carbon cycles, e.g. photosynthesis or water absorption.

Slow carbon cycle

This 'slow' cycle refers to the cycling of carbon between rock stores and the atmosphere and oceans through the processes of weathering over millions of years. Weathering of rocks on continents creates a net carbon sink in the oceans. Chemical weathering of rocks by carbonic acid produces carbonate runoff, which is transferred to the oceans. Here, organisms use it to create shells and when the organisms die the carbonate sediment produced eventually forms limestone. This long-term carbon store is released to the atmosphere through volcanic activity.

Figure 1.14 The subsystems of the carbon cycle

TESTED

26 Explain the difference between the slow and the fast carbon cycles.

27 How is carbon dioxide removed from the atmosphere?

Answers on p. 259

The carbon cycle and scale

The carbon cycle can be studied at a range of scales. For example, the terrestrial carbon cycle can be studied at the scale of:

+ an individual plant (for example, a tree)
+ a field
+ a local ecosystem or
+ a continent.

Figure 1.15 shows the carbon cycle at the individual plant scale.

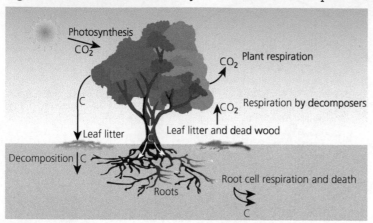

Figure 1.15 Carbon cycle of a single tree

Processes driving the transfer of carbon

Table 1.4 outlines the processes driving the transfer of carbon.

Table 1.4 Processes driving the transfer of carbon

Process	Description
Photosynthesis	+ Plants use energy from the sunlight and combine CO_2 from the atmosphere with water from the soil to form carbohydrates. + Virtually all organic matter is formed from this process. Carbon is stored (or sequestered) for long periods of time as trees can live for hundreds or thousands of years and resistant structures such as wood take a long time to decompose.
Respiration	+ Plants release CO_2 back into the atmosphere due to respiration (about half of the terrestrial portion). + In soil respiration, microscopic organisms living in soil also release CO_2 through respiration.
Decomposition	+ The process of decomposition by fungi and bacteria returns CO_2 to the atmosphere. + Decomposition also produces soluble organic compounds dissolved in runoff from the land surface. + Decomposition releases greenhouse gases (GHGs) as a by-product.
Combustion	+ Fossil fuels (coal, oil and natural gas) contain carbon captured by living organisms over periods of millions of years and stored in the Earth's crust. + Since the Industrial Revolution, fossil fuels have been mined and combusted as a primary energy source. + CO_2 is the main by-product of fossil fuel combustion.

Exam tip

You must be clear on the processes for each topic; however, it is also important to understand the factors that affect the rate of those processes. For example, increased sunlight will lead to increased photosynthesis and consequently greater uptake of water and CO_2 by plants. The revision activity below will help you with this.

Ecosystem A community of living organisms, their relationship to each other and the environment.

Greenhouse gas Any gas in the atmosphere that allows short-wave UV solar radiation to pass through into the atmosphere but prevents terrestrial infrared radiation from escaping into space.

Check your understanding and progress at **www.hoddereducation.co.uk/myrevisionnotesdownloads**

Process	Description
Carbon sequestration in oceans and sediments + oceanic carbon pump + biological pump	+ CO_2 moves from the atmosphere to the ocean by the process of diffusion. + At low latitudes, warm water absorbs CO_2. + At high latitudes where cold water sinks, the carbon is transferred deep into the ocean. + Where the cold water returns to the surface and warms again, it loses CO_2 to the atmosphere. + In this way CO_2 is in constant exchange between the oceans and the atmosphere. + This vertical circulation is a process called the 'oceanic carbon pump'. + Phytoplankton also fix CO_2 through photosynthesis – the carbon passes through the oceanic food web. Shell-building organisms remove carbonate from the sea. + When organisms die, the shells sink into deep water; decay of marine organisms releases some CO_2 into the deep water (the biological pump). + Some material forms layers of carbon-rich sediments which, over millions of years, turn to sediments in rocks.
Weathering	+ Weathering processes (driven by the atmosphere, rain and groundwater) break down rocks on the Earth's surface. + These small, weathered particles combine with plant and soil particles and are eventually carried to the ocean. + Large particles are deposited on the shore. The sediment accumulates. + Layers build up and eventually, due to surface pressure, form shale rock. + Within the ocean, dissolved sediments mix with the seawater. Marine organisms use the dissolved sediments to make skeletons and shells containing calcium carbonate. When these organisms die, the carbonate collects at the bottom of the ocean and sedimentary rocks (for example, limestone) form.

Now test yourself

TESTED ⦿

28 What is carbon sequestration?

29 How can warm and cold climates affect decomposition rates on land?

Answers on p. 259

Changes in the carbon cycle over time

Wildfires

+ Plant carbon enters the atmosphere in the event of a wildfire.
+ Wildfires:
    + remove dense areas of carbon-storing plants
    + eliminate plants that would take CO_2 out of the atmosphere as they grow
    + expose soil, which releases carbon from decaying plant matter.
+ Young plants, crops or alternative land uses that store less carbon replace the burned vegetation, so there is a net decrease in the carbon store.
+ All of these effects increase the CO_2 in the atmosphere.
+ As CO_2 is the most important gas for controlling the Earth's temperature, increased CO_2 accelerates the process of greenhouse heating.

Carbon sequestration
The process of capturing and storing CO_2.

Weathering The breakdown of rocks in situ by a combination of weather, plants and animals.

Revision activity

Draw a cross-section of the ocean and annotate it with information about the functions of the oceanic carbon pump and the biological pump.

Revision activity

Draw a topic web/spider diagram which summarises the potential impacts of wildfires on the carbon cycle. Colour code the diagram into short-term and long-term impacts. Try to go beyond the text on this page (e.g. link impacts to the greenhouse effect).

Volcanic activity

The full impact of volcanic activity remains uncertain. However, Figure 1.16 summarises, in a cause-and-effect sequence, the potential impact of volcanic explosions on the carbon cycle.

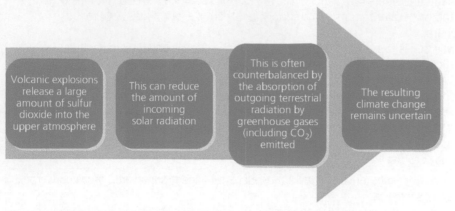

Figure 1.16 The impact of volcanic activity on the carbon cycle

Farming practices

+ Ploughing introduces air into the soil, decomposition accelerates and carbon is released to the atmosphere.
+ Emissions from tractors increase the level of CO_2 in the atmosphere.
+ Livestock release methane gas to the atmosphere as a by-product of digestion.
+ Rice paddies generate methane.

Land use change: deforestation

Figure 1.17 shows how the flow of carbon in a tropical forest can change from a carbon sink to a carbon source as a result of deforestation.

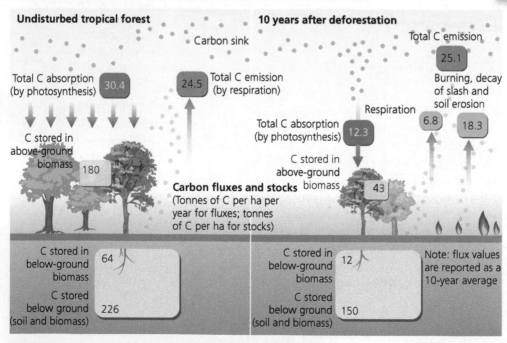

Figure 1.17 The effects of deforestation on the carbon cycle

Carbon sink A carbon store that absorbs more carbon than it releases.

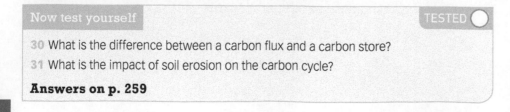

Check your understanding and progress at **www.hoddereducation.co.uk/myrevisionnotesdownloads**

Land use change: urban growth
+ Urban growth reduces the amount of surface vegetation.
+ CO_2 emissions from energy consumption, transport, industry and domestic use increase.
+ There is an increase in CO_2 emissions from cement manufacture required for more building.

Carbon sequestration is the capture of carbon from the atmosphere and placing it in long-term storage. It can be achieved in the following ways:
+ Through geological sequestration, where CO_2 is captured at its source and injected in liquid form deep underground.
+ In oceans, as they absorb CO_2, it sinks into deep ocean stores within weeks and circulates for thousands of years.
+ Using plants to capture and store CO_2 in stems, roots and soil, known as biological sequestration.

> **Now test yourself** TESTED ⊙
>
> 32 What measures can help mitigate the impact of urban growth on the carbon cycle?
>
> 33 What is biological sequestration?
>
> **Answers on p. 259**

> **Revision activity**
>
> Use one of the following methods – bullet point notes, flow diagram (Figure 1.16) or picture diagram (Figure 1.17) – to explain how the extraction and burning of hydrocarbon fuel can change the carbon cycle.

Carbon budgets
The Earth's carbon cycle is in a constant state of motion. The processes described above are constantly transferring carbon between stores, working over a full range of timescales from seconds to millennia.

If the carbon moving into any given pool is the same as the carbon being transferred out of that pool, the system is in a state of **dynamic equilibrium**. A **carbon budget** is a list of all the carbon pools with an estimate of their size and a summary of all the fluxes that constitute inputs and outputs.

> **Carbon pool** Carbon stores.

The Earth's carbon budget at present shows that it is a system in imbalance, mainly due to:
+ fossil fuel combustion
+ land use change.

The impact of the changing carbon cycle

Land
+ The amount of carbon that plants take from the atmosphere has risen in recent years.
+ With more atmospheric CO_2 to convert to plant matter through the process of photosynthesis, plants have been able to grow more – this is carbon fertilisation.
+ Plants will continue to grow and absorb CO_2 until they reach a limit in the amount of water or nitrogen available.
+ Wildfires are generally extinguished, preventing large amounts of carbon from entering the atmosphere from this source.
+ In some parts of the world, such as New Zealand, China and parts of Europe, more intensive agriculture has allowed some farmland to return to more dense vegetation, which can store more carbon.
+ In other parts of the world where temperatures are high, dry trees are more susceptible to fire and forests may burn more readily, releasing CO_2, for example the Australian bushfire season in 2019–2020.
+ In areas of water scarcity, trees:
  + slow their growth and take up less carbon, or
  + die and release their stored carbon into the atmosphere.

Oceans

+ Ocean acidification – dissolving CO_2 in the ocean creates carbonic acid, which increases the acidity of water. Carbonic acid reacts with carbonate ions.
+ Over time, ocean acidification results in fewer carbonate ions in seawater so shell-building animals, such as corals, have thinner and more fragile shells.
+ Coral reefs are threatened and there is a fall in marine biodiversity.
+ Warmer oceans are a product of the greenhouse effect. Phytoplankton grows better in cool, nutrient-rich waters and so the ocean's ability to take carbon from the atmosphere could reduce.
+ CO_2 is also essential for the growth of phytoplankton and an increase in CO_2 could increase the growth of some species.
+ Ocean warming kills algae, which corals need to grow, leading to bleaching and eventual death of reefs.
+ When sea ice melts, the reflective ice is replaced by more heat-absorbent water; the ocean absorbs more sunlight, which amplifies the warming process.
+ Ocean salinity is decreasing in the North Atlantic, probably due to a knock-on effect from:
    + higher levels of precipitation which, through runoff, eventually enters the sea
    + the melting of ice sheets, which also adds fresh water to the sea.
+ Such changes impact the circulation of the North Atlantic waters and eventually impact the climate of north-west Europe.
+ Melting terrestrial ice and thermal expansion are causing global sea levels to rise – rates of 3.5 mm per year since the early 1990s have been recorded. This is a eustatic rise in sea level.

Atmosphere

+ CO_2 is the most important gas for controlling the Earth's temperature.
+ Scientists have calculated that CO_2 causes about 20% of the Earth's greenhouse effect, water vapour about 50% and clouds 25%. Aerosols and methane cause the rest.
+ While CO_2 contributes less to the overall greenhouse effect, it is the gas that sets the temperature, so CO_2 controls the amount of water vapour in the atmosphere and thus the size of the greenhouse effect (see Figure 1.18).

> ### Making links
>
> Links can be made between the water and carbon cycles at all scales from global to local, for example how climate change is putting pressure on the water cycle (page 19) as well as on the carbon cycle (page 31).

> ### Now test yourself
> TESTED ◯
>
> 34 What is carbon fertilisation?
>
> **Answer on p. 259**

> ### Exam tip
>
> It is important to note that all of these impacts of a changing carbon cycle involve gains and losses. They are complex in nature and it is difficult to predict the precise rate, magnitude and direction of change. There will also always be exceptions.

> ### Revision activity
>
> After revising 'the impact of the changing carbon cycle', either draw a spider diagram or create a table that categorises the impact by type, i.e. economic, political, social and physical.

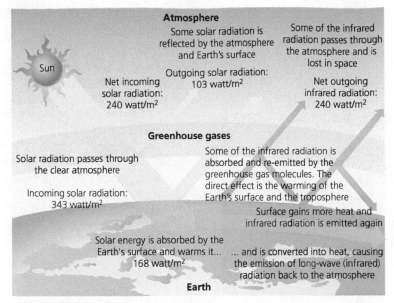

Figure 1.18 The greenhouse effect

The **enhanced greenhouse effect** is the amplification of the greenhouse effect due to increased amounts of CO_2 and other greenhouse gases in the Earth's atmosphere as a result of human activities. The result is a further increase in global temperatures.

Now test yourself — TESTED ○

35 What is the enhanced greenhouse effect as opposed to the greenhouse effect?

Answer on p. 259

Revision activity

This is a peer group activity. From memory, draw an annotated diagram of the greenhouse effect and evaluate each other's attempts. Think about accuracy of content, clarity and appropriate sequence.

Revision activity

Organise the labels in Figure 1.18 into a numbered sequence to help you revise and understand the functioning of the greenhouse effect.

Water, carbon, climate and life on Earth

REVISED ○

The key role of the carbon and water stores and cycles supporting life on Earth

At a global scale, water and carbon flow in closed systems between the atmosphere, lithosphere, biosphere and oceans.

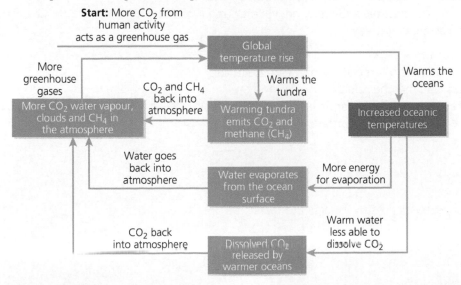

Figure 1.19 The link between the carbon cycle, the water cycle and the atmosphere

Making links

It is important to understand how the carbon and water cycle are linked.

The role of feedbacks

Although CO_2 contributes less to the overall greenhouse effect than water vapour, CO_2 sets the temperature as it affects radiation in the atmosphere and causes an overall warming effect. The warming affects the amount of water vapour in the atmosphere (see Figure 1.19).

Climatic feedbacks can affect global warming as either positive or negative feedback (see Figure 1.20).

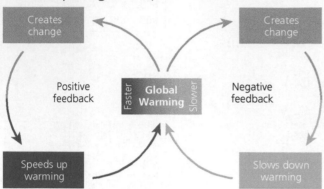

Figure 1.20 Climatic feedbacks

Exam tip

Remember to practise interpreting graphs and data related to the role of feedbacks. You may be asked to interpret composite graphs showing changes over time (e.g. changes in CO_2 or temperature) or scatter graphs showing the relationship between two factors (e.g. temperature and water vapour).

Climatic feedbacks can have a significant impact on the magnitude of potential future climate change.

+ Water vapour feedback: an increase in CO_2 leads to an increase in the temperature of the atmosphere; warm air holds more water vapour; water vapour helps the Earth hold on to more heat energy from the Sun = warmer climate. **Positive feedback**.
+ Albedo feedback: bright surfaces; increased reflection of sunlight; warmer climate; snow melt; less sunlight reflected; more warming = warmer climate. **Positive feedback**.
+ Clouds: not fully understood. Clouds reflect sunlight back to space, leading to cooling, but clouds can also restrict heat radiated back, leading to warming. The key is in the fact that high clouds retain heat for longer, while low cloud is thicker and reflects more sunlight. In a warming world we may get more high cloud. **Some positives and some negatives**.
+ Terrestrial carbon cycle: some CO_2 in the atmosphere is absorbed by plants and the soil, some by oceans. Changes to land surfaces in the future will affect CO_2 release and absorption, for example soils may be warmer and release CO_2. **Some positives and some negatives**.

The following feedbacks are uncertain:

+ Feedback of methane hydrates – there is a large stock of methane in the deep ocean; in a warming ocean this may be released.
+ Permafrost areas – carbon locked in organic-rich soils in permafrost areas in high latitudes may be released in a warming climate.

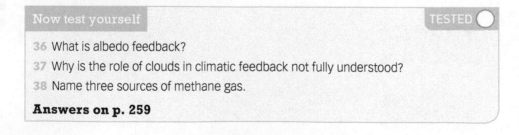

Now test yourself   TESTED ⚪

36 What is albedo feedback?
37 Why is the role of clouds in climatic feedback not fully understood?
38 Name three sources of methane gas.

Answers on p. 259

Check your understanding and progress at **www.hoddereducation.co.uk/myrevisionnotesdownloads**

Human interventions in the carbon cycle designed to influence carbon transfers and mitigate the impacts of climate change

Mitigation refers to any method used to reduce or prevent the emission of greenhouse gases.

Strategies to mitigate greenhouse gas emissions

Carbon capture and sequestration technologies (CCS)

CCS aims to capture large percentages of the CO_2 emissions from the burning of fossil fuels. There are three parts to the CCS system:

+ capture
+ transportation
+ storage.

Changing rural land use

+ Grasslands: soil carbon storage in grasslands can be improved by:
    + adding minerals and fertilisers to increase organic matter and increase plant productivity, which will absorb more CO_2 from the atmosphere
    + irrigation and water management, which will improve plant productivity, as will revegetation with improved pasture species.

+ Croplands: soil carbon storage can also be improved on land used to grow crops.
    + Mulching adds organic matter and prevents carbon losses from the system.
    + Rotation of cash crops with cover crops can increase the biomass returned to the soil.
    + Improved crop varieties can increase productivity and enhance soil organic carbon (SOC).
    + Forests reduce CO_2 emissions to the atmosphere by storing carbon above and below ground and absorbing carbon from the atmosphere. Therefore, forest protection, reforestation and agroforestry are all important.

Transport innovations

Attempts to reduce greenhouse gas emissions from road and aviation transport form a key element of mitigation.

Road transport initiatives include sustainable transport schemes, congestion charging (London), park-and-ride schemes (Cambridge), integrated transport networks (Curitiba, Brazil).

CO_2 migration within the aviation industry includes:

+ movement management (for example adopting fuel-efficient routes)
+ flight management (for example cruising at a lower speed)
+ design technology (for example carbon capture within engines).

> **Exam tips**
>
> There is not always direction in an exam question to refer to case studies, but remember that it is always good exam technique to support your ideas with case study evidence where appropriate, and especially in extended writing responses
>
> The water and carbon cycles are studied separately, but there are interactions and inter-relationships between the two cycles, particularly through processes. For example, life-supporting processes such as photosynthesis use CO_2 and water.

> **Exam tip**
>
> There is a wide range of strategies for mitigating greenhouse gas emissions. Use these strategies and categories of scale to structure an essay. It is also important to evaluate these strategies – address their positives and negatives and possible futures.

> **Revision activity**
>
> Create a table with four columns listing greenhouse gas emission mitigation attempts at different scales: local, regional, national and global.

> **Soil organic carbon** The organic components of soil, e.g. tissues from dead plants and animals.

My Revision Notes: AQA A-level Geography Second Edition

Case studies

A tropical rainforest setting, for example the Amazon Basin

This case study should:

+ illustrate and analyse the key themes in the water and carbon cycle
+ link these key themes to environmental change and human activity in the rainforest.

What do I need to know?	Content and suggested revision methods
Key themes of the carbon cycle + Stores + Transfers Key themes of the water cycle + Stores + Transfers	Organise your notes into a table of bullet pointed notes, e.g.: Place-specific facts should be highlighted and learnt – their use will access the higher marks.
How human activity has driven change in the rainforest	Organise your notes into a spider diagram. The centre bubble of the diagram should be 'How human activity has driven change in the rainforest'. Each 'stem' should be an example, e.g.: + deforestation + pollution + slash-and-burn agriculture + replacing forest with pastureland. For each example of human activity also include details of the **impacts on the carbon and water cycle**. For example, for 'Replacing forest with pastureland': forest land absorbs approx. 11% more solar radiation than pasture land.
The effects of environmental change	You can subdivide the effects of environmental change into effects on: + climate + vegetation + soil + rivers. Produce a simple diagram that summarises the impacts on each of the above. For example, for 'Environmental effects on soil': When forests are cleared and burned, 30–60% of the carbon is lost to the atmosphere; soil bacteria that recycle dead vegetation die. Carbon store in **soils** decreases.

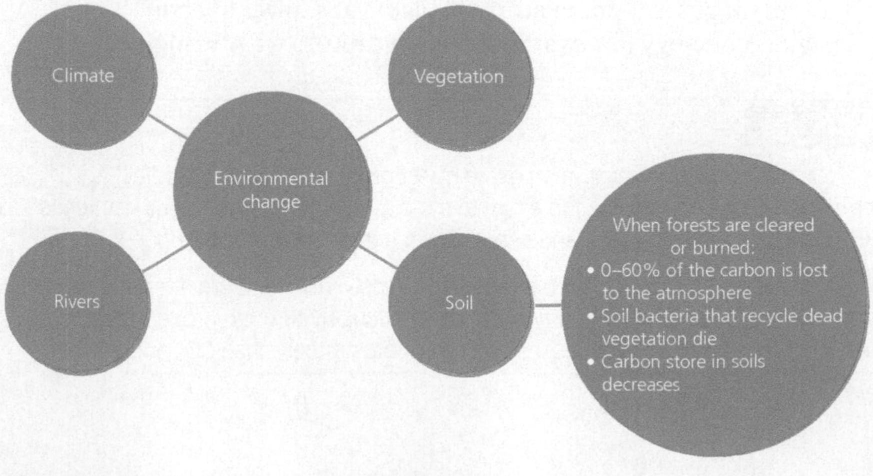

Nested table within first row:

Carbon cycle	Water cycle
Stores: + Estimated store of 76 million tonnes of carbon in Amazonia in 2019.	Stores: + In the north-west part of the Amazon Basin average rainfall can exceed 6,000 mm annually.
Transfers: + Growth spurt of trees in the Amazon Basin due to more CO_2 intake.	Transfers: + 48% of evapotranspiration falls again as rain.

Figure 1.21 Effects of environmental change

Check your understanding and progress at www.hoddereducation.co.uk/myrevisionnotesdownloads

What do I need to know?	Content and suggested revision methods
Mitigation strategies to reduce environmental impacts	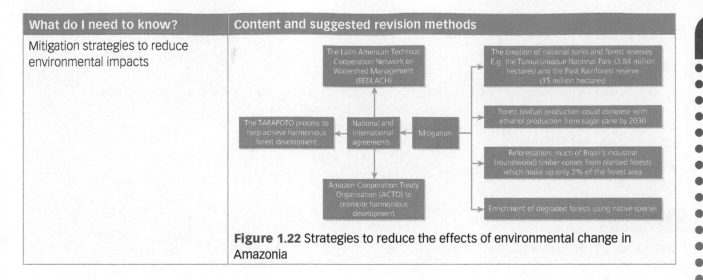 **Figure 1.22** Strategies to reduce the effects of environmental change in Amazonia

A river catchment at a local scale

This case study should:

+ illustrate and analyse the key themes in the water cycle
+ engage with fieldwork data
+ consider the impact of precipitation on drainage basin stores and transfers
+ consider the implications for sustainable water supply and/or flooding.

What do I need to know?	Content and suggested revision methods
The geographical context of the chosen river catchment	You can summarise this in an annotated map. Produce a sketch map with annotations that provide information on the catchment area: + general topography + types of land use (forest land, National Park land, farmland, settlements, transport routes) + the river course and sizeable tributaries.
Precipitation data and the impact on drainage basin stores and transfers	You can summarise this through a series of annotated graphs and charts which show information such as rainfall totals and river flow/discharge over a period of time (e.g. a year). Your annotations should explain the **impacts** of precipitation on stores and transfers within the river catchment. The data may be from primary/fieldwork or secondary sources.
The implications of the stores and transfers of flow within the catchment for flooding (or water supply)	Revision notes for this section could be in the form of bulleted notes. You could also create a table, with one column for information/facts regarding flood events or water supply sources for example and another column to consider the implications for flood defences or ensuring sustainable water supplies.

> **Exam tip**
>
> Making links includes the ability to apply your knowledge and understanding to unfamiliar locations and examples. Remember, when analysing data, do not get thrown just because you do not recognise the location.

33

Exam practice

1 Outline **two** of the processes that transfer carbon from one pool to another. [4]

2 Explain the climatic feedback between water vapour and climate change. [4]

3 Outline **two** challenges associated with safe levels of groundwater abstraction. [6]

4 Assess the need for land-use planning in flood risk areas. [6]

5 Examine the importance of forest trees in the carbon cycle. [6]

Answers and quick quizzes online

Exam skills

Opportunities to practise geographical skills within this topic include:

+ analysis of specific graphs such as:
  + flood hydrographs
  + soil budget graphs
  + line graphs showing seasonal changes in water storage and surface runoff

+ geospatial data, i.e. global maps showing:
  + oceanic warming and cooling
  + oceanic circulation flows
  + global levels of forestry.

Summary

+ Be clear on the systems approach and the concepts of positive and negative feedback and dynamic equilibrium.
+ For both the carbon and water cycles understand the meaning of the lithosphere, hydrosphere, cryosphere, atmosphere and biosphere, the major stores of carbon and water, their size and geographical distribution.
+ Key processes affect the flows and transfers of both water (evaporation, condensation, cryospheric processes) and carbon (photosynthesis, respiration, decomposition, combustion, carbon sequestration and weathering).
+ The cycling of water exists at the global, drainage basin and slope scale. There are a number of common inputs, outputs, stores and flows.
+ Be clear on the concepts of water balance and carbon budgets and the factors affecting them.

+ Be able to analyse and interpret hydrographs showing river regimes and storm responses.
+ Natural and human factors lead to changes in the water and carbon cycles over time.
+ Understand the impacts of changes on the water and carbon cycles – these may be economic, social or environmental.
+ There are key links between the water and carbon cycles and the atmosphere – linkage of knowledge is a key feature of A-level geography.
+ A combination of initiatives is needed to mitigate the impact of climate change.
+ It is important to develop a view/opinion on possible futures.

Check your understanding and progress at **www.hoddereducation.co.uk/myrevisionnotesdownloads**

2 Hot desert systems and landscapes

Deserts as natural systems

Systems in physical geography

The **systems** approach is a way of analysing the relationships within a unit, for example a hot desert environment. It consists of a number of components and the linkages between them, which can be represented in a flow diagram.

A hot desert landscape is an **open system**. It has:
+ **inputs** (energy – insolation, water, wind and sediment), stores/components (characteristic erosional and depositional desert landforms)
+ **flows/transfers** (by agents of wind and water)
+ **outputs** of both energy (clear skies allow energy to be re-radiated back to space) and matter, which cross the boundary of the system to the surrounding environment. (See Figure 2.1.)

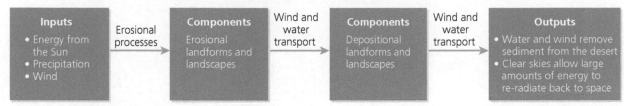

Figure 2.1 The hot desert landscape as an open system

Hot deserts are dynamic (constantly changing) places:
+ The system is in a state of **dynamic equilibrium**.
+ Change occurs to upset the balance of the system – for a hot desert, this may be due to desertification (see Figure 2.2).
+ The system adjusts by a process of **feedback**, which can be either:
    + **positive** (progressively greater change from the original condition of the system) or
    + **negative** (the system is returned to its original conditions).

> **Desertification**
> Land degradation in dryland areas due to overexploitation by humans and natural processes such as drought.

> **Revision activity**
>
> Produce a diagram for a negative feedback in a hot desert environment. Figure 2.2 gives an example of a positive feedback.

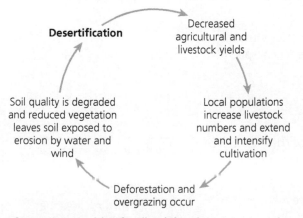

Figure 2.2 Positive feedback in a hot desert environment – desertification

Now test yourself

TESTED ◯

1 Give two examples of inputs to a desert system.
2 Why are hot deserts considered to be open systems?
3 Distinguish between positive and negative feedback adjustment.
4 Define dynamic equilibrium.

Answers on p. 259

Exam tip

When asked to distinguish between two terms, such as positive and negative feedback, state the precise meaning of each term and then draw out the difference(s).

Making links

The systems approach is widely used in physical geography. See Chapter 1 page 9 for more information on systems.

The global distribution of mid- and low-latitude deserts

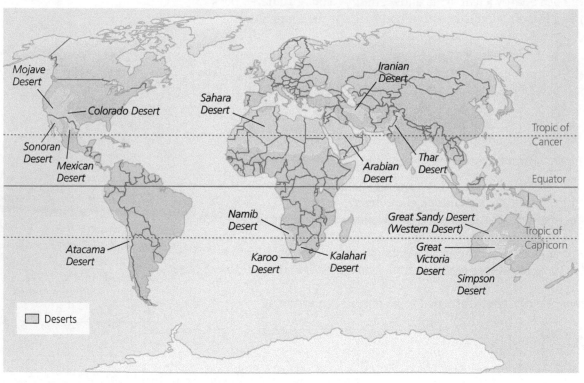

Figure 2.3 The location of major hot desert environments

Revision activity

Describe and account for the distribution of deserts shown in Figure 2.3. Remember to:
+ use geographical vocabulary, such as 'in the northern or southern hemisphere'
+ use the names of major continents
+ be specific about lines of latitude
+ use compass points, for example north, south.

Hot deserts generally run in parallel belts north and south of the equator in hot arid and semi-arid, mid- and low-latitude locations (see Figure 2.3).

The main hot deserts are:
+ Mojave, Sonoran, Colorado and Mexican deserts in the western part of North America
+ Sahara, Thar and Arabian deserts on the Tropic of Cancer
+ Atacama, Namib, Kalahari, Great Sandy and Simpson deserts on the Tropic of Capricorn.

Arid An area that receives less than 250 mm of precipitation per year.

Semi-arid An area that receives between 250 mm and 500 mm of precipitation per year.

Check your understanding and progress at **www.hoddereducation.co.uk/myrevisionnotesdownloads**

Making links

For more on the causes and types of precipitation, see Chapter 1 page 13.

Exam tip

If you are presented with an exam question with data for an unfamiliar desert location, with wording such as 'Using Figure xx and your own knowledge …', remember, it is the application of your knowledge and understanding of deserts to that location that is important rather than specific knowledge about that particular desert.

Characteristics of hot desert environments and their margins

Climate
+ Extremes characterise desert climate.
+ Wide annual temperature range (35°C summer and 10°C winter, on average).
+ Wide diurnal temperature range (30°C to 0°C on average).
+ Clear skies lead to rapid heat loss at night and high levels of insolation during the day.
+ Low humidity.
+ Desert margins have a wide climatic variation as seasonality comes into effect.
+ Low rainfall (generally < 250 mm annually) – rain comes in the form of unpredictable, intense cloud bursts.
+ Annual rainfall can be as low as 15 mm, as in the Atacama Desert, Chile.

Now test yourself   TESTED ◯

5 Describe and explain the diurnal temperature range in hot deserts.
6 Why is desert rainfall often in the form of intense cloud bursts and what is the name for this type of rainfall?

Answers on p. 259

Soils
+ Arid soils – aridisols.
+ The two main categories within the aridisols are:
  + **sierozems** (these form in desert – shrub areas where there is a little more vegetation)
  + **raw mineral soils** (where physical and chemical weathering forms a coarse-textured soil).
+ Slow rates of weathering and lack of vegetation mean that soils in hot deserts are shallow.
+ They are unproductive – there are minerals and nutrients there but limited vegetation.
+ They have a tendency to be saline (evaporating moisture leaves salts behind) and alkaline.
+ Calcium is concentrated near the surface due to capillary action where moisture in the soil moves upwards.

Aridisols Soils which form in arid or semi-arid climates.

Exam tip

Rainfall **effectiveness** is important in arid areas. There is rainfall which quickly evaporates before it can become effective.

Vegetation
+ Many desert plants survive by reducing water loss by transpiration.
+ Among the water-saving strategies are shedding leaves, small leaves, leaves whose stomata close during the day, leaves covered in a thick, waxy cuticle. Further adaptations of plants to drought and salinity are shown in Table 2.1.

Table 2.1 Adaptations of plants to drought and salinity

Plant type	Adaptation
Succulents	Plants that store water within their tissues (e.g. prickly pear).
Phreatophytes	Plants with long roots to tap water deep below the surface (e.g. tamarisk).
Drought evading	Annual plants that germinate and set seed when it rains. The seeds remain dormant until the next rain (e.g. desert paintbrush).
Dormant	Perennial plants that lie dormant during dry spells and spring to life only when water becomes available.
Halophytes	Plants adapted to survival where salt concentrations are high and toxic to most species (e.g. saltbush, which stores fresh water in its fleshy leaves).

7 Why are desert soils a) thin and b) saline?

8 Outline three adaptations of desert plants to the climatic conditions.

Answers on p. 259

Water balance and aridity index

✛ Water balance compares the mean annual precipitation (P) received with the mean annual potential evapotranspiration (**PET**).

✛ An aridity index is the ratio of P and PET (a numerical indicator of the degree of dryness of the climate at a given location) – see Table 2.2.

Table 2.2 The aridity index

Classification	Aridity index (AI)	Global land area (million km²)	Global land area (%)
Hyper-arid	< 0.05	10.0	7.5
Arid	0.05 to < 0.20	16.2	12.1
Semi-arid	0.20 to < 0.50	23.7	17.7

> **Making links**
>
> For more on water balance, see Chapter 1 page 15.

The causes of aridity

Atmospheric circulation

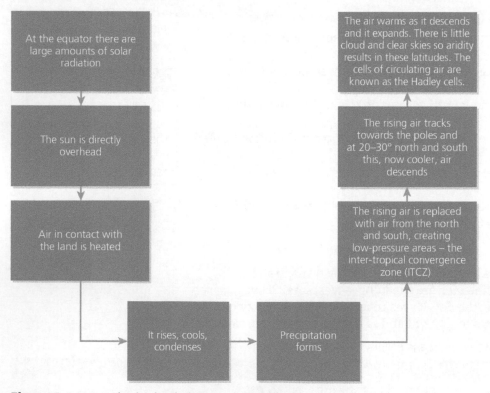

At the equator there are large amounts of solar radiation

The sun is directly overhead

Air in contact with the land is heated

It rises, cools, condenses

Precipitation forms

The rising air is replaced with air from the north and south, creating low-pressure areas – the inter-tropical convergence zone (ITCZ)

The rising air tracks towards the poles and at 20–30° north and south this, now cooler, air descends

The air warms as it descends and it expands. There is little cloud and clear skies so aridity results in these latitudes. The cells of circulating air are known as the Hadley cells.

Figure 2.4 Atmospheric circulation

> **Exam tip**
>
> Remember to offer a clear **cause and effect** sequence of explanation to answers based on the causes of aridity.

Distance from the sea

Distance from the sea can lead to temperature extremes. Inland areas some distance from the sea are generally much drier as land heats quickly during the day due to the lack of clouds inland (see Figure 2.5).

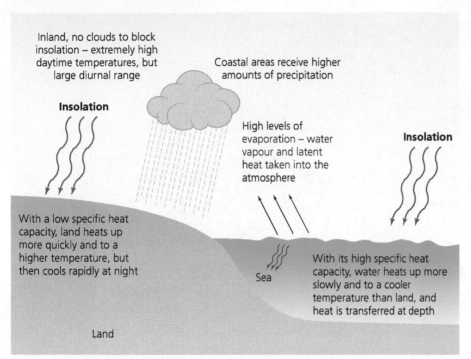

Figure 2.5 The effect of continentality

Revision activity

Practise simple flow diagrams to explain the influence of the following factors on aridity:
+ atmospheric circulation
+ **continentality**
+ relief
+ cold ocean currents.

Continentality The effect on the climate of increased distance from the sea.

Relief

Areas on the leeward side of mountains lie in a **rain shadow area**, for example central Australia to the west of the Eastern Ranges. Reasons for this are:
+ Air rising over the mountains cools, condenses and expends much of its rainfall.
+ Air descending on the downward side warms and the relative humidity is lowered, reducing the rainfall, for example the Kalahari Desert lying in the rain shadow of the Drakensberg Mountains in South Africa.

Now test yourself   TESTED ◯

9 What is the aridity index?

10 What are Hadley cells?

11 Why do temperatures vary so much between day and night in inland locations with no cloud cover?

12 Define what is meant by a rain shadow area.

Answers on p. 259

Cold ocean currents
+ Wind moving over cold water is cooled, relative humidity increases and moisture condenses to form fog and mist offshore.
+ The land heats more quickly than the sea, creating low pressure over the land as warm air rises.
+ An onshore wind (blowing from high to low pressure) blows the fog and mist inland and it is burned off by the heat of the overhead sun.
+ Stable, sinking air means there is little cloud formation and no rain, for example the influence of the Peruvian current on the Atacama Desert.

Exam tip

A combination of the factors discussed under 'The causes of aridity' can often cause hot deserts to form. Be specific by relating the factors to a named example.

Systems and processes

Sources of energy in hot desert environments

Energy sources are shown in Figure 2.6.

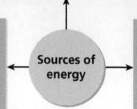

Insolation
- Changes in temperature drive processes
- Heat origin – (approx.) 12 hours of sunshine a day
- Sun is high, giving a high angle of incidence, so radiation is concentrated on a small surface
- Moisture enables latent heat to escape as water evaporates
- As there is little moisture, there is no cooling effect of the escape of latent heat

Runoff
- Rainfall is localised and unpredictable in terms of time and space
- Ground conditions of baked earth act like concrete and there is limited vegetation ⇒ all increase overland flow (runoff), which becomes an agent of erosion and transportation

Sources of energy

Wind
- Localised winds, e.g. harmattan of the Sahara
- Winds blow outwards towards the edge of the desert
- Wind acts as an agent of erosion and transportation

Figure 2.6 Sources of energy in hot desert environments

Now test yourself

TESTED

13 Define the term 'insolation'.

14 Outline the weathering processes that break down parent material.

Answers on p. 259

Exam tip

Questions which ask you to **outline** something require an accurate description.

Making links

In Figure 2.5 runoff is referred to as an energy source. See Chapter 1 page 19 for more information on flows in the water cycle and how ground conditions can affect processes.

Sediment sources, cells and budgets

Sediment sources

These come from:
+ weathering of parent material
+ fluvial sources – rivers – if they are ephemeral (they dry up), the sediment is left behind on dry riverbeds
+ aeolian – wind-blown deposits – loess.

Aeolian Relating to wind action.

Sediment cells and budgets
+ Areas dominated by **erosion** are a source of sediment for other areas and their system has a **net sediment loss**.
+ Areas dominated by **deposition** receive sediment and their system has a **net sediment gain**.

Exam tip

You could be asked about the **relative importance** of different processes in the formation of specific desert landforms. Read the question very carefully for the specific focus and make sure that you assess the importance of the different processes involved in the formation of a feature rather than just describing how it is formed.

Revision activity

Make a copy of Figure 2.6 on A3 paper and add information on deposition and erosion processes (e.g. aeolian erosion processes, such as deflation, abrasion and attrition).

Geomorphological processes

Geomorphological processes are outlined in Figure 2.7.

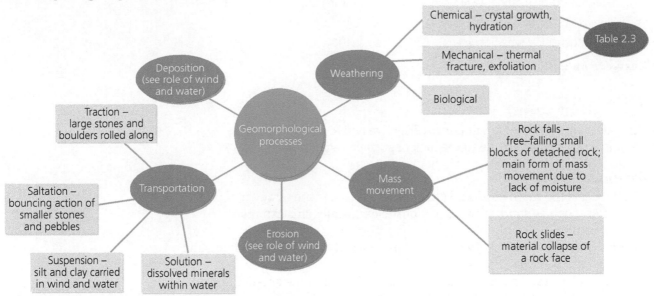

Figure 2.7 Geomorphological processes in hot desert environments

> **Exam tip**
>
> Be clear on the difference between **weathering** processes and **erosion** processes.
> + Weathering is the in-situ breakdown of exposed rock by physical, chemical or biological processes.
> + Erosion involves the wearing away and removal of particles by wind and water.

Distinctively arid geomorphological processes

Weathering refers to the main processes forming desert landscapes due to:
+ regular heating and cooling of surfaces
+ the presence of even small amounts of moisture.
+ the presence of living organisms (see Table 2.3).

Table 2.3 Distinctively arid geomorphological processes

Process	Description
Thermal fracture (**mechanical weathering**)	High diurnal temperature ranges cause expansion and contraction in rock, leading to disintegration over a long period of time.
Exfoliation (mechanical weathering)	'Onion skin weathering': + As weathering and erosion take place at the surface, pressure is released from rocks at depth. + Cracks form, running parallel to the surface. + Capillary action brings salts to the surface. + The salts deposited in cracks and enhanced chemical weathering peel layers of rock from the surface.
Crystal growth (chemical weathering)	+ Water present in joints and bedding planes evaporates, leaving salts behind. + The salt crystals grow over time, exerting pressure. + Heating and cooling lead to expansion and contraction, which assist in the physical breakdown of the rock.
Hydration (chemical weathering)	Absorption of even the smallest amount of moisture (such as dew) causes rock to swell, making it vulnerable to further mechanical breakdown.
Block and granular disintegration	+ The processes above can lead to breakdown of rock into large blocks, where bedding planes and joints are prominent. + As mechanical and chemical processes take effect, **block disintegration** occurs. + Where individual grains are broken away from rock surfaces by the effects of thermal expansion and contraction or freeze–thaw action of moisture, **granular disintegration** occurs.

15 What three factors contribute to weathering processes in desert environments?

16 Distinguish between block and granular disintegration.

17 What is mechanical weathering?

Answers on p. 259

The role of wind

In hot desert environments wind is an effective active agent of erosion, transport and deposition due to the lack of surface vegetation cover.

Erosion

+ **Deflation** is the main erosional effect of wind. This is the removal of fine particles, such as sand, silt and clay particles. Locally, this can result in dust storms.
+ **Abrasion** is the scouring effect of sand particles blown by the wind on solid rock. This is a slower process than deflation.
+ **Attrition** also takes place as grains of sand collide, become smaller and rounder and can be used effectively.

Transportation

The main methods of transportation by wind are:
+ **creep** – sand grains slide and roll across the surface
+ **saltation** – the skipping motion of sand grains as they are lifted and fall again
+ **suspension** – very small dust particles carried in the air.

Deposition

When the power of wind falls below a critical erosion velocity, transportation ceases and deposition occurs. Sand dunes form.

Now test yourself TESTED

18 Give an example of an erosional process that is active in a desert environment.

19 What is the difference between wind erosion and wind transportation?

20 What specific flows would contribute to desert **sheet flow** (an overland flow of water where the water is not concentrated into channels)?

Answers on p. 260

> **Exam tip**
>
> There is enough moisture available in hot deserts to allow a range of weathering processes to take place. Even small amounts of dew are enough for example. Do not overlook the importance of weathering processes in an explanation of landforms. They weaken rock and allow erosion to be more effective.

Sources of water

Water forms landforms in hot deserts. Table 2.4 summarises the types of water source.

Table 2.4 Hot desert water sources

Water source	Description
Exogenous	Perennial rivers, which flow in hot desert environments but gain their water in humid regions, for example Colorado.
Endoreic	River catchments within hot desert environments. They are often seasonal.
Ephemeral	Short-lived water sources that flow only after heavy rain. When they dry up, they are known as wadis.

The episodic role of water

+ Sheet floods result when there is an intense downpour that quickly runs off hard-baked surfaces, which are impermeable. Overland flow intensifies.
+ This movement of water can become an effective erosional force, creating landscape features and contributing to deposition elsewhere.
+ When the water is concentrated down a wadi (often steep-sided and narrow), flash flooding occurs.

> ## Making links
>
> See Chapter 1 page 17 for more on flash flooding hydrographs. You should be able to apply the concepts to hot desert environments – for example, what is it about ground conditions and type of rainfall in deserts that leads to flash flooding?

> ## Revision activity
>
> 1 Make a revision table of weathering and erosion processes. Pay particular attention to the relative role of each in the formation of different landforms in a desert landscape. Your table should have three columns:
>   a Name of process
>   b Description
>   c Landforms that the process produces
> 2 Practise your exam technique by explaining the processes at work in the desert landscape shown in Figure 2.8. Look at the photograph carefully. It shows a large boulder with the surface 'peeling' away, lots of large, angular blocks of rock and pieces of rock of varying size and shape.
> 3 On A3 paper, make a summary of the main features that make up a 'desert landscape'. This could be an annotated sketch or a spider diagram or table.
>   a Name each of the features
>   b For each feature, give a short, one-sentence description of how it looks and list the processes leading to its formation. For example:
>   Canyon: a narrow valley with almost vertical sides. Formed by vertical river erosion and mass movement processes such as mudflows and landslides.

> ## Exam tip
>
> When referring to processes in landform formation, be specific. For example, if you mention weathering, specifically what type of weathering? This is all part of giving depth and detail to an answer.

Figure 2.8 Death Valley, California

Now test yourself   TESTED ⭕

21 What are the sources of moisture in a desert?

22 What is an ephemeral stream?

Answers on p. 260

Arid landscape development in contrasting settings

REVISED ⭕

Origin and development of landforms of mid- and low-latitude deserts

Aeolian landforms

Table 2.5 summarises the main aeolian landforms.

Table 2.5 Aeolian landforms

Feature	Description	Formation
Deflation hollows	Wind removes dry sand, silt and clay.	Where the material is 'scooped' out, it leaves a deflation hollow.
Desert pavements	A surface layer of closely packed gravel and pebbles.	As fine material is removed, coarse material and pebbles are left behind, forming a desert pavement.
Ventifacts	Exposed rocks lying on the desert surface.	These are shaped by the abrasion (sand-blasting effect) of wind-blown sediment.
Yardangs	Streamlined parallel ridges of rock, aligned in the prevailing wind direction.	Caused by wind-blown erosion processes – abrasion.
Zeugen	The collective term for rock pillars, rock pedestals and yardangs.	Less resistant underlying rock has been eroded to leave more resistant rock at the surface.

Now test yourself   TESTED ⭕

23 How are ventifacts formed?

24 Outline the role of wind in the formation of barchans.

25 Explain how the star dune pictured in Figure 2.9 is formed.

Answers on p. 259

Figure 2.9 A star dune

> **Exam tip**
>
> When examining landscape features, you will sometimes be asked to respond to a photograph so it is important to practise examining photographic evidence. For example, you may be asked to name a feature in the photograph.

44

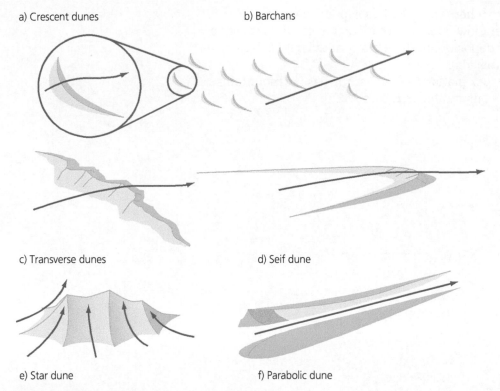

a) Crescent dunes

b) Barchans

c) Transverse dunes

d) Seif dune

e) Star dune

f) Parabolic dune

Figure 2.10 Dune types

Four main types of dune form, as shown in Figure 2.10:

+ **Crescent dunes:** wider than they are long with a concave slip face. There are two types of crescent dunes:
  + **barchans** – mounds and ridges of blown sand (with horns that face downwind). To form, they require an adequate supply of sand, strong and frequent wind, and an obstacle to trap the sediment. Creep and saltation transport the sand up the windward slope, sand accumulates on the peak and eventually a small avalanche will occur down the slip face to restore equilibrium. In this way, dunes advance in the direction of the prevailing wind. Figure 2.11 shows the formation of a dune
  + **transverse dunes** – a feature of larger environments, they can be several hundred metres high.
+ **Seif dunes:** linear dunes, straight or slightly curved. Often more than 100 km long and can be over 200 m high, with a slip face on alternate sides, they cover large areas in parallel, knife-edged ridges.
+ **Star dunes:** pyramidal in profile, with slip faces on three or more sides. They form where the wind comes from different directions.
+ **Parabolic dunes:** have a 'U'-shaped form with arms that extend upwind.

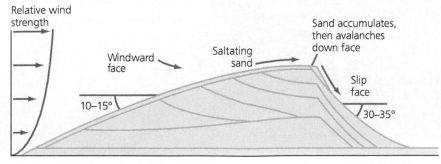

Relative wind strength

Windward face

Saltating sand

Sand accumulates, then avalanches down face

Slip face

10–15°

30–35°

Figure 2.11 Formation of desert dunes

Water landforms

+ **Wadis:** steep-sided, wide-bottomed, gorge-like valleys formed by fluvial erosion. They are rarely filled with water and have a build-up of sediment on the valley floor. They are either permanently dry or they have ephemeral streams, which are fast-flowing and the result of intense storms.

+ **Bahadas:** alluvial fans form where rivers leave steep-sided valleys (canyons) and enter adjacent lowlands. The reduction in gradient causes a sudden loss of energy and deposition. Where many alluvial fans develop it is referred to as a bahada (bajada).
+ **Pediments:** gently sloping rock platforms found at the base of mountains in hot desert environments, as illustrated in Figure 2.12.

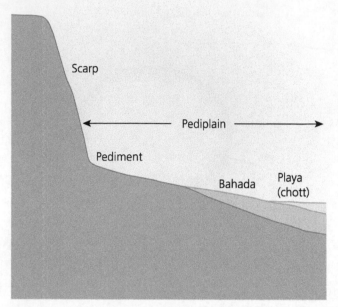

Figure 2.12 Location of a pediment

+ **Playas:** where ephemeral streams flow into inland basins, salt lakes or playas form. These are often temporary and where the water has evaporated, salts are left behind. Sodium chloride, the most common, can give the appearance of a beach.
+ **Inselbergs:** rounded, steep-sided hills that rise suddenly from a lowland plain. They are generally composed of solid crystalline rocks such as granite and are thought to be relics of previous geomorphological processes.

Exam tip

Remember that in the past many desert environments had a more humid climate than today and so water had a past influence on desert landscapes, which can still be seen today.

Revision activity

Remember that it is important to support your answers with examples. Either as an additional annotation to one of the tables or diagrams you have already prepared or as a separate list/table, name a located example for the main features of a desert landscape. For example, Death Valley in California contains named examples of many landforms, and the Namib Desert in south-west Africa has examples of different dune types.

Now test yourself   TESTED ◯

26 What are the main aeolian erosion processes in desert environments?

Answer on p. 260

Revision activity

Practise a clear sequence of explanation for the formation of three erosional and three depositional landforms.

Check your understanding and progress at **www.hoddereducation.co.uk/myrevisionnotesdownloads**

The relationship between process, time, landforms and landscapes

Figure 2.13 summarises the factors of process, time, landform and landscape.

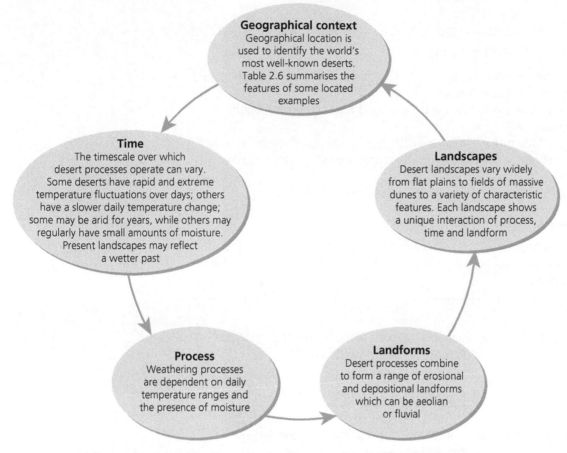

Figure 2.13 Process, time, landforms and landscapes in hot desert environments

---

Making links

See Chapter 1 for more information on fluvial processes. For example, surface/overland flow is high in desert landscapes due to sparse vegetation and baked earth, whereas infiltration is low due to hard surfaces and throughflow is low due to shallow soils.

Deserts have their own unique combination of landforms to produce distinctive landscapes. Table 2.6 summarises some of the landscapes of the most well-known deserts.

Exam tip

Be clear on whether examination questions refer to landforms or landscapes.

Table 2.6 Physical features of different desert landscapes

Desert	Characteristic landscape features
Arabian (Arabian Peninsula)	Almost entirely sandy, with some of the largest sand dune systems in the world
Australian Great Sandy, Simpson and Great Victoria Gibson and Sturt	Mostly sandy plains Mostly stony surfaces
Chihuahuan (Arizona, New Mexico, Texas and north central Mexico)	High, flat plateau with some stony surfaces and sandy soil, broken by mountain ranges and distinctive mesas
Kalahari (south-western Africa)	Extensive sand dunes interspersed with gravel plains
Mojave (Arizona, California, Nevada)	A varied landscape including sandy soils, gravel pavements and salt flats
Sahara (northern Africa)	Vast ranges of dunes among mountains and rocky areas, and gravel plains and salt flats
Thar (India and Pakistan)	Mostly sand dunes with areas of gravel plains

Desertification

UNESCO gives the definition of desertification as 'the persistent degradation of dryland ecosystems by human activities and by climate change'.

The changing extent and distribution of hot deserts over the last 10,000 years

The extent of hot deserts has changed with the climatic changes in the glacial and interglacial periods of the Pleistocene.

+ During the last glacial period, the extent of arid areas (therefore, cold deserts) was vast.
+ 8,000 years ago, during the 'Holocene Climate Optimum', the extent of hot deserts was confined to relatively small areas, mainly in north Africa and the Middle East.
+ Present-day deserts extend over much larger areas.

Areas at risk from desertification

The UN reports that approximately 1 billion people are at further risk from desertification and that around 12 million hectares of land annually are lost due to desertification.

> **Exam tips**
>
> In exam answers, represent a balanced understanding of the causes of desertification and an appreciation of the interaction of human and physical factors.
>
> Desertification is a complex process, which can lead to a variety of impacts. These are not always 'desert like' and can refer to areas with accelerated soil erosion.

Figure 2.14 shows the extent of areas at risk from desertification today.

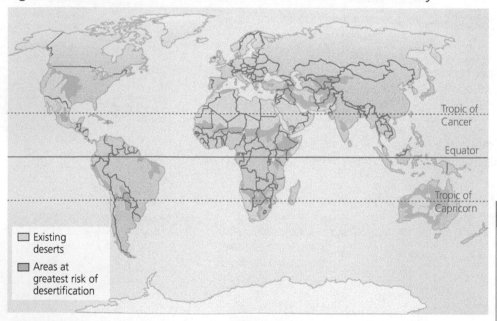

☐ Existing deserts

☐ Areas at greatest risk of desertification

Figure 2.14 Areas at risk from desertification

The causes of desertification

Desertification now ranks among the greatest environmental challenges of our time, with 168 countries worldwide and 15 billion people affected.

The situation is acute in countries such as Somalia, Ethiopia, Djibouti and Kenya, where the combination of lack of government focus and prolonged periods of drought linked to climate change is driving desertification levels. Figure 2.15 summarises the causes of desertification.

> **Revision activity**
>
> Using the map in Figure 2.14, identify two countries:
> + one for which desertification presents a severe challenge (think about links to poverty and low levels of food security)
> + one that reflects the impact of desertification in an advanced country.
>
> Make a list of points to explain why desertification occurs and what impact it has on the named country.

Check your understanding and progress at **www.hoddereducation.co.uk/myrevisionnotesdownloads**

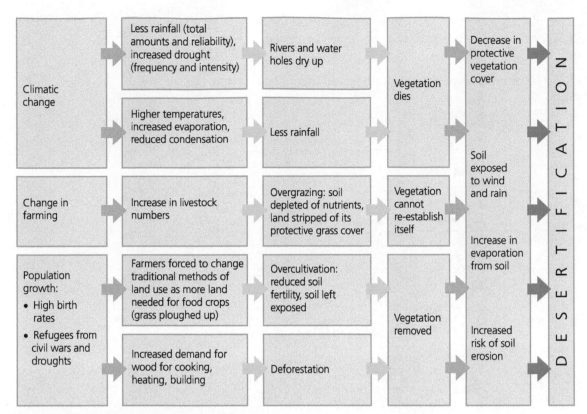

Figure 2.15 The causes of desertification

Impact of desertification

Desertification impacts ecosystems, landscape and populations (see Table 2.7).

Table 2.7 Impacts of desertification

Impact on ecosystems	Impacts on landscape	Impacts on populations
+ Reduction in vegetation, leading to reduced habitats and increased competition + Continual cropping decreases nutrient recycling + Soil nutrients are lost through wind and water erosion + Carbon sinks are reduced + Loss of biodiversity + Food webs become fragile	+ Increased erosion + Increased number of sand dunes + Increased sedimentation of rivers + More sand storms + Vegetation damaged by 'sandblasting' winds	+ Dryland populations are often socially and politically marginalised due to poverty and remoteness + Forced migration + Increased male outmigration + Loss of traditional knowledge and skills + Reduced availability of fuel wood leads to increased purchase of kerosene, with health issues + Food shortages + Reduced income from traditional economy + Widespread rural poverty

Now test yourself　　　　　　　　　　　　　　　　TESTED ◯

27　What is desertification?

28　Explain how population growth can lead to desertification.

29　State three impacts of desertification on ecosystems.

30　How is desertification linked to poverty?

Answers on p. 260

Predicted climate change and the potential impact

With some scientists suggesting that global temperatures rose by between 0.3°C and 0.6°C during the twentieth century, and that they could increase further by between 2°C and 5°C in the twenty-first century, there is little doubt about the potential for further increases in degraded land and deserts.

The problem of desertification is complex and has multiple causes and possible solutions. It will continue to affect some of the most vulnerable populations.

> **Exam tip**
>
> Credit will be given for pointing out that famine rarely means an absolute food shortage for everyone. It often impacts the poorest section of society with the lowest level of food security.

Solutions range from global responses to climate change to small-scale, appropriate technology and 'bottom-up' approaches to aid local communities.

Figure 2.16 shows the feedback links between desertification, climate change and biodiversity loss.

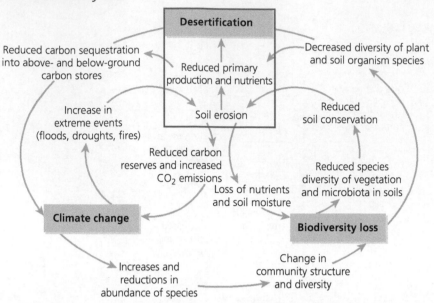

Figure 2.16 Links and feedback between desertification, climate change and biodiversity loss

Revision activity

Make a list of global and local measures to prevent future desertification.

Alternative possible futures for local populations

The key points on desertification and the future are as follows:

+ Poverty and poor agricultural practices will continue to be the main cause of desertification.
+ Relieving the human pressure on 'at risk' areas is key.
+ Food insecurity, deforestation and land degradation are linked issues that cause desertification.
+ Climate is predicted to become not only warmer but also less predictable and more extreme, leading to increases in droughts and floods.
+ Local populations need direct help in the form of a variety of solutions that are proactive, make use of local knowledge and improve their resilience.

The following points summarise possible measures to prevent future desertification and restore degraded land. Intervention is aimed at both the local and the global level:

+ Improve agricultural practices.
+ Invest in integrated land and water management techniques to protect soils and reduce over-grazing.
+ Support science-driven agriculture through, for example drought-resistant crops.
+ Maintain vegetation cover to protect soils and re-establish soil fertility through the use of organic fertilisers.
+ Reduce clearance of shrubs and trees by developing non-wood energy supplies, which are naturally available in the dryland ecosystem, for example solar, biogas and wind power.

Exam tip

Possible futures is an important focus of the course. Make sure you are aware of how desertification can be addressed at the local and the global level.

Now test yourself

33 Describe two methods that can be used to reverse the effects of desertification.

34 Why are desert environments described as 'fragile'?

Answers on p. 260

TESTED ⬤

Case studies

A hot desert environment setting, for example fieldwork measuring aeolian processes that shape dunes in a local coastal sand dune environment

This case study should:
+ illustrate and analyse the key themes in the hot deserts unit
+ engage with field data.

What do I need to know?	Content and suggested revision methods
How aeolian processes have shaped sand dunes For example, the effect of wind speed and direction on sediment movement and sand dune profile on a stretch of sand dunes	You could organise your revision notes into a series of annotated graphs and charts. The annotations will provide information on the prevailing wind direction and wind speed across a section of sand dunes. An annotated sketch diagram of the dune profile could summarise the sand dune (i.e. its slope angle, vegetation cover and shape). You could create a revision summary in the form of bulleted points to outline the relationship between sand dune shape (e.g. steepness and height) and aeolian processes.

A landscape at a local scale where desertification has occurred, for example Touat region, southern Algeria

This case study should:
+ illustrate and analyse the key themes of desertification (the causes and impacts of desertification and the implications for sustainable development)
+ evaluate human responses of resilience, mitigation and adaptation.

What do I need to know?	Content and suggested revision methods
Causes and impacts of desertification	It is important to revise a clear cause-and-effect sequence of explanations. Flow diagrams are a good way to explain each cause and its impacts/effects. For example: Population increases (2% per annum in Algeria) → Increased demand for water for cultivation → Intensive water supply projects (e.g. modern water pumps extracting huge quantities of water from groundwater stores) → Water stores are depleted, leading to dry land and desertification
Implications of desertification for sustainable development	Relate the implications to the causes of desertification that you have outlined. So, in the example above, water supply becomes **less sustainable** as deeper wells are dug to tap even deeper groundwater supplies. Implications can be a bulleted list in revision notes.
How people have been able to adapt to and mitigate the effects of desertification	In addition to factual information outlining the adaptations and mitigations, it is important to **evaluate** these responses (the advantages and disadvantages, how successful the measures have been/are). A table, like the one below, detailing the responses and evaluating them is a useful way of organising your revision notes for this section. **Mitigation and adaptations** / **Evaluation** Schemes to preserve traditional irrigation methods (foggaras) (Add a simple diagram to explain how these work) / Advantages: + Water flows underground, preventing evaporation and water loss + Building materials are locally sourced, providing work + Powerful diesel pumps are not used Disadvantages: + 'Water deciders' hold responsibility for distributing water, requiring a high degree of trust and fairness.

2 Hot desert systems and landscapes

1   Outline the sources of moisture in hot desert environments.   [4]
2   Explain the geomorphic processes most influential in forming the landform shown in Figure 2.18.   [4]

Figure 2.18 A sand dune in the Sahara, southern Morocco

3   Explain how hot deserts can be viewed as a system.   [6]
4   With reference to Figure 2.19, explain why some areas of the world are more vulnerable to desertification than others.   [6]

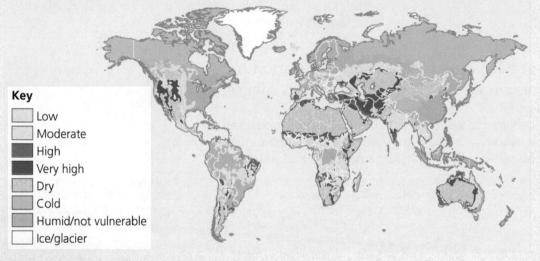

Key
- Low
- Moderate
- High
- Very high
- Dry
- Cold
- Humid/not vulnerable
- Ice/glacier

Figure 2.19 Vulnerability to desertification

5   Explain the relationship between process, time and landforms in the creation of hot desert landscapes.   [6]

Answers and quick quizzes online

Exam skills

Opportunities to practise geographical skills within this topic include:

+ observation skills in fieldwork
+ collection, manipulation, presentation and analysis of primary and secondary fieldwork data, including quantitative and qualitative sources (particularly in relation to the first case study)
+ analysis of specific graphs such as climate graphs and hydrographs
+ observation/interpretation of photographs
+ interpretation of geospatial data (for example global maps showing desertification, soil types, climate data).

Summary

+ The systems approach can be applied to the study of desert environments. Positive and negative feedbacks can lead to environmental issues such as desertification, which is the result of a positive feedback.
+ It is the interaction of rainfall and evapotranspiration that explains the formation of desert environments, not rainfall alone.
+ Vegetation in desert environments has developed a range of adaptations. Understand how desert climate, vegetation and soils are interconnected.
+ Simple diagrams are an effective way of explaining the cwtauses of aridity: atmospheric circulation, continentality, relief and cold ocean currents.
+ There is a range of weathering processes at work to loosen rock prior to the work of erosion processes.
+ The moisture available in desert environments plays a significant role in weathering and erosion processes and landform development.
+ Desertification affects both rich and poor countries.
+ It is important to balance explanation of desertification in terms of the physical and human causes.
+ Case studies and examples should allow effective evaluation of the responses to desertification.

Check your understanding and progress at **www.hoddereducation.co.uk/myrevisionnotesdownloads**

3 Coastal systems and landscapes

Coasts as natural systems

REVISED

Systems in physical geography

The **systems** approach is a way of analysing the relationships within a unit, for example, a coast. It consists of a number of components and the linkages between them represented in a flow diagram, as in Figure 3.1.

The coast is an **open system**. It has:

+ **inputs** (energy in the form of waves, wind, tides and currents)
+ **stores/components** (beach and sediment accumulation), **flows/transfers** (movement of sediment, for example by processes such as longshore drift)
+ **outputs** of both energy and matter (marine and wind erosion from beaches and rock surfaces and evaporation), which cross the boundary of the system to the surrounding environment.

The combination of all these factors forms distinctive landscapes, which are made up of a range of erosional and depositional landforms and reflect human activity.

Coasts are dynamic (constantly changing) places:

+ The system is in a state of **dynamic equilibrium** with a balance between inputs and outputs.
+ Change occurs to upset the balance of the system – for a coast this may be due to landslides, storms or human activity, for example.
+ The system adjusts by a process of **feedback**, which can be either:
  + **positive** (progressively greater change from the original condition of the system) or
  + **negative** (the system is returned to its original condition).

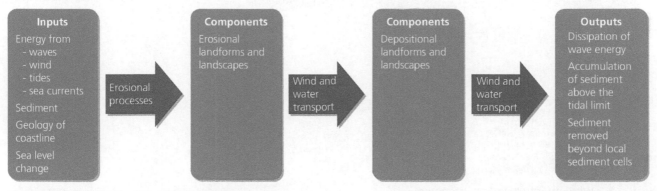

Figure 3.1 The coast as an open system

TESTED

Now test yourself

1 Why are coasts classified as open systems?
2 Define dynamic equilibrium.
3 How do coastal systems get their energy?
4 How does geology affect the coastal system?

Answers on p. 260

Revision activity

Add the following labels to a copy of Figure 3.1:

+ Thermal energy from the sun
+ Materials from marine deposition
+ Mass movement on cliffs
+ Evaporation

My Revision Notes: AQA A-level Geography Second Edition

Making links

The systems approach is widely used in physical geography. See Chapter 1 page 9 for more information.

Exam tip

When asked to distinguish between two terms, such as positive and negative feedback, state the precise meaning of each term and then state the difference between them.

For example, positive and negative feedback – both are adjustments to a system. Positive refers to a progressive greater change from the original condition of the system and negative is when the system is restored to its original condition.

Revision activity

Produce a diagram which explains how the coastal system responds to a positive and a negative feedback, for example, strong winds. Think about the impact on erosion, the production of sediment and how this sediment could be used in further erosion or be deposited.

Systems and processes

REVISED

Sources of energy in coastal environments

Wind

Wind is the primary source of energy for a range of other processes such as erosion, transportation and deposition (see Figure 3.2). These aeolian processes contribute to the shaping of coastal landscapes.

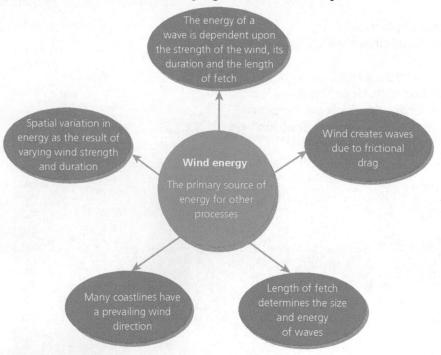

Figure 3.2 Wind energy in coastal systems

Waves

Waves are undulations on the surface of the sea driven by wind.

Key features:
+ Height – the difference between the crest and the trough of a wave.
+ Length – the distance between crests.
+ Frequency – the time lapse between crests.

Functioning:
+ A wave enters shallow water → friction with the seabed increases → the wave slows → increases in height → and plunges or breaks onto the shoreline.
+ The wash of water up the beach is the swash; the drag back down the beach is the backwash.

Check your understanding and progress at **www.hoddereducation.co.uk/myrevisionnotesdownloads**

Wave types: see Table 3.1.

Table 3.1 Wave types

Wave type	Characteristics
Constructive	✤ Low, long wavelength (up to 100m) ✤ Low frequency (6–8 per minute) ✤ Gentle spill onto the shore ✤ The swash loses volume and momentum, leading to a weak backwash so sediment movement off the beach is low ✤ Material is slowly and gradually moved up the beach, forming berms
Destructive	✤ High, steep wavelength ✤ High frequency (10–14 per minute) ✤ Rapid approach to shoreline and wave steepens rapidly before breaking ✤ Little forward movement of the water creates a powerful backwash so sediment is pulled away from the beach ✤ Very little material is moved up the beach ✤ Forms storm beaches

Exam tip

Be clear on the use of the terms 'constructive waves' and 'destructive waves'. Do not assume that destructive waves create erosion features and constructive waves create depositional features. The description of the waves refers to the transportation of sediment in the coastal zone, that is constructive waves deposit sediment and destructive waves remove it.

Now test yourself   TESTED ◯

5 What is a prevailing wind and how will it affect a specific coastline?

6 What is meant by the term 'length of fetch'?

7 What factors affect the energy of a wave?

Answers on p. 260

Revision activity

Draw a simplified diagram of a constructive wave and a destructive wave, and apply the descriptions given in the text above. Use the outlines in Figure 3.3. Add as much detail as you can using your own notes and pages 55–56.

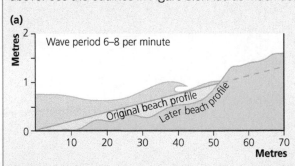
(a)

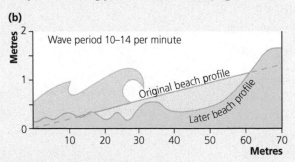
(b)

Figure 3.3 Outline diagrams for **(a)** a constructive wave and **(b)** a destructive wave

Exam tip

In physical geography, learning the correct sequence to an explanation is key to achieving accuracy.

Now test yourself   TESTED ◯

8   Explain the process of wave refraction on headlands and bays.

Answer on p. 260

Wave refraction

Wave refraction is the process by which waves break onto an irregularly shaped coastline, such as a headland separated by two bays.

Waves drag in the shallow water approaching a headland → the wave becomes high, steep and short → the part of the wave in the deeper water moves forward faster → the wave bends → the low-energy wave spills into the bays as most of the wave energy is concentrated on the headland.

Currents

Currents are the permanent or seasonal movement of water in the seas and oceans. There are three types, as outlined in Table 3.2.

Table 3.2 Types of current

Current type	Characteristics and effects
Longshore	Most waves approach the shoreline at an angle. This creates a current of water running parallel to the shoreline. Effect: transports sediment parallel to the shoreline.
Rip	These are strong currents moving away from the shoreline due to a build-up of seawater and energy along the coastline. Effect: hazardous for swimmers.
Upwelling	The global pattern of currents circulating in the oceans can cause deep, cold water to move towards the surface, displacing the warmer surface water. Effect: a cold current rich in nutrients.

Now test yourself   TESTED ◯

9   Distinguish between ocean waves and ocean currents.

10 Which type of ocean current plays an important part in the transportation of coastal sediment?

11 Why do waves break?

Answers on p. 260

Tides

See the revision activity below.

Revision activity

The notes above on wind, waves and currents illustrate three different methods in making revision notes:
+ a spider diagram
+ notes with bullet points and bold or colour print for key terms
+ a table summary.

Choose one of these methods to make your own revision summary of the final source of energy in coastal systems – **tides**. Include the following key terms:
+ spring tide
+ neap tide
+ tidal range
+ tidal/storm surges.

Making links

See Chapter 1 page 11 for more on the circulation of water between oceans.

Tides The rise and fall in sea level in response to the gravitational pull of the Sun and the Moon.

Check your understanding and progress at **www.hoddereducation.co.uk/myrevisionnotesdownloads**

Low-energy and high-energy coasts

Figure 3.4 summarises the various features of both high-energy and low-energy coastlines.

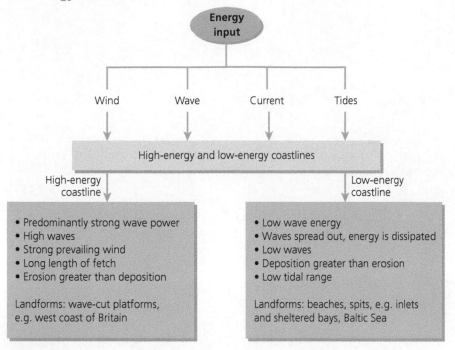

Figure 3.4 The features of high- and low-energy coastlines

> **Exam tip**
>
> Be very clear on the dominant processes for different locations – for example, tidal energy dominates in an estuarine environment while dune environments tend to be dominated by wind action. Take into account the geographical context of a location/example. When examination resources state, 'Use Figure x and your own knowledge ...', you don't need to know about the specific location used in the question. Rather, it is the transfer and application of your knowledge to an unfamiliar location that is important.

Sediment sources, cells and budgets

Sediment sources

The crustal sediments that form depositional features, such as beaches and mudflats, originate from the following sources:

+ **Seabed:** rising sea levels over the past 18,000 years have meant that sediment from continental shelf areas has been swept towards the shoreline.
+ **Rivers:** rivers account for 90 per cent of coastal sediment, a combination of bedload shingle and suspended silt and clay.
+ **Cliff erosion:** sediment from erosion contributes only 5 per cent or less to coastal systems.
+ **Biological origin:** for example, shells and corals. Again, this makes up a very small percentage of the sediment budget.

Note that rivers and seabed sources account for the highest proportion of sediment sources, not cliff erosion as is often thought.

Sediment cells

Sediment movement occurs in distinct areas called cells (see Figure 3.5). If part of a larger cell, they are called sub-cells. For example, the Flamborough Head–Humber Estuary sub-cell is part of the larger Flamborough Head–The Wash cell.

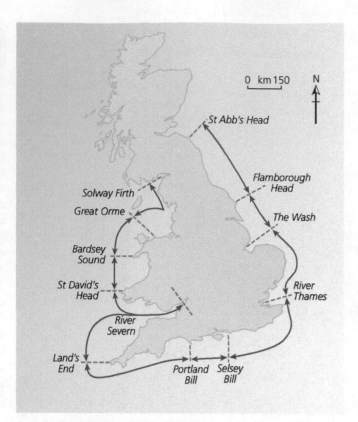

Figure 3.5 Coastal sediment cells around England and Wales

Sediment budgets

The **sediment budget** is the balance between sediment being added to and removed from the coastal system.

+ Positive budget = more material added than removed (accretion of material) → shoreline builds to the sea.
+ Negative budget = more material removed than added → shoreline recedes landward.

Calculating sediment budgets is complex as all possible inputs, stores (sinks) and outputs of sediment need to be identified.

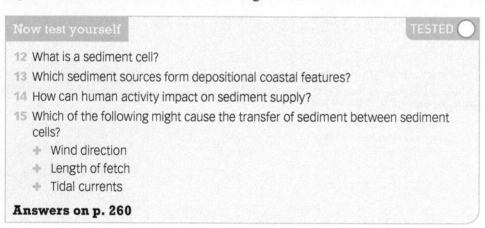

Revision activity

Make a table of two columns, one side for a positive sediment budget and one for a negative sediment budget. In each column, list the ways of managing such coastlines.

Making links

See more on coastal management on pages 69–71 of this chapter.

Making links

See Chapter 1 pages 15–16 for more on water balance and soil water budgets.

Coastal processes

Coastal landforms develop as a result of a range of interrelated climatic and geomorphic processes:

+ **Weathering** processes alter rock by chemical and mechanical breakdown. These weathering processes also supply material used in erosion processes.
+ **Mass movement**, marine processes and **aeolian processes** also contribute to landform formations.

Marine processes
Processes connected with the sea operating upon a coastline, for example waves and tides.

Figure 3.6 shows the range of coastal processes, while Table 3.3 summarises these processes.

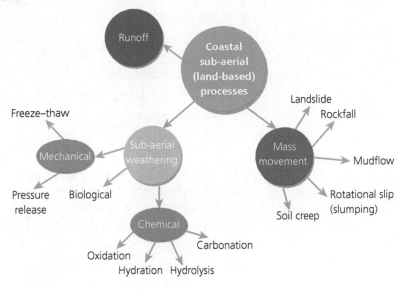

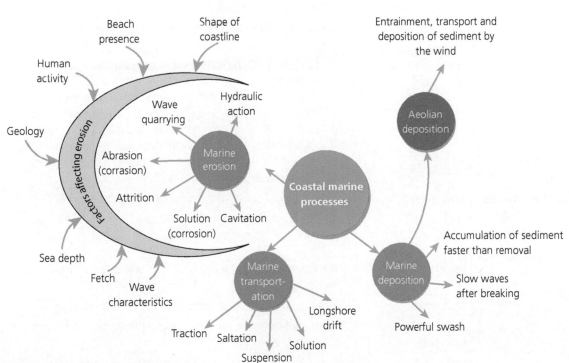

Figure 3.6 Coastal processes

Table 3.3 Key coastal processes and descriptions

Group of processes	Name of process	Description
Marine erosion – connected with the sea action	Hydraulic action	Wave pounding: the force of the water on the rocks
	Wave quarrying (cavitation)	Breaking wave traps air in cracks in a cliff face; as the water pulls back, air is released under pressure, which weakens the rock face over time
	Abrasion/corrasion	Sand, shingle and boulders picked up by the sea and hurled against a cliff
	Attrition	The wearing down of rocks and pebbles as they rub against each other, making them smaller and rounder
	Solution/corrosion	Where fresh water mixes with salt water, acidity may increase and carbon-based rocks at the coast will be broken down
Marine transportation	Traction	Large boulders roll along the seabed
	Saltation	Small stones bounce along the seabed
	Suspension	Very small particles are carried in moving water
	Solution	Dissolved material
	Longshore drift	Waves approach the shore at an angle, swash moves material up the beach in the same direction as the wave, backwash moves the material back down the steepest gradient – usually perpendicular – where it is picked up by the next incoming wave
Marine and aeolian deposition		Occurs on low-energy coastlines or where there is an abundance of erosion material
Sub-aerial (operate on the land) weathering processes	Mechanical weathering	Climate related, e.g. freeze–thaw weathering; pressure release of underlying rock – where overlying material is removed by erosion, weathering or mass movement
	Biological weathering	Break down by the action of vegetation and other coastal organisms
	Chemical weathering	+ Oxidation – O_2 dissolved in water reacts with some rock minerals, e.g. iron-rich rocks + Hydration – physical addition of water to minerals in rocks makes them more susceptible to chemical weathering + Hydrolysis – mildly acidic water reacts with minerals + Carbonation – CO_2 dissolved in rainwater makes a weak carbonic acid, which reacts with calcium carbonate in limestone and chalk
Sub-aerial mass movements – dependent upon slope angle, grain size, temperature and saturation	Landslides	Cliffs made of softer rocks slip when lubricated by rainfall
	Rockfalls	Rocks undercut by the sea or slopes affected by mechanical weathering
	Mud flows	Heavy rain causes fine material to move downhill
	Rotational slip/ slumping	Where soft material overlies resistant material and excessive lubrication takes place
	Soil creep	Very slow movement of soil particles down slope
	Runoff	The movement of water across the hard surface, carrying debris

Exam tip
It is unlikely that a question will simply ask you to explain how a particular feature is formed. You are more likely to be asked about the **relative importance** of different processes in the formation of named coastal landforms. When assessing the importance of different processes in the formation of a feature, it is important to understand and explain the geographical context, i.e. climate, prevailing wind, temperature range, amount of precipitation, importance of mechanical and chemical weathering.

Exam tip
It does not matter which term you use – abrasion/ corrasion or solution/ corrosion – best practice is to pick one, learn it and stick to it.

Check your understanding and progress at **www.hoddereducation.co.uk/myrevisionnotesdownloads**

Making links

See Chapter 1 page 16 and think about how processes in the drainage basin system affect coastal areas – as rivers enter the sea.

Now test yourself

TESTED ○

16 How would each of the following factors affect coastal erosion?
  + Length of fetch
  + Wave characteristics
  + Beach presence

17 What is the difference between weathering and mass movement, and how does mass movement contribute to landscape development?

Answers on pp. 260–61

Revision activity

In pairs, and using Table 3.3, test each other on the key processes. Either give the term and ask for an explanation or give the explanation and ask for the term to which it refers.

Coastal landscape development

REVISED ○

Coastal landscapes and the associated landforms reflect the interaction of a range of factors and processes.

+ Geology:
  + Coastal configuration – headlands attract energy due to wave refraction.
  + Rock characteristics (lithology) – resistant rocks such as chalk erode more slowly than weak rocks such as clay; structure – cracks and fissures can be exploited; concordant and discordant coastlines – lines running parallel (concordant) or at right angles (discordant).
+ Climate:
  + Temperature ranges lead to more freeze–thaw weathering, wet climates lead to slope failure.
+ Nature of tides and waves:
  + Wave steepness: steeper waves = high energy.
  + Length of fetch: long fetch = more powerful waves.
  + Sea depth: steep shelving will create higher, steeper waves.
+ **High- and low-energy input:** high-energy waves = more erosion.
+ **Human activity** and **coastal management**.

Fetch The distance of open water over which wind blows without being interrupted by land obstacles.

There have been attempts to classify coasts based on the factors that affect them and the landforms that characterise them. The main categories are:
+ concordant and discordant
+ cliffed, flat or graded shoreline
+ **emergent** (features are exposed as sea level falls) and **submergent** (flooded as sea level rises).

Now test yourself

TESTED ○

18 What is the difference between hydraulic action and wave quarrying?

19 Name three types of chemical weathering.

20 How does climate affect mass movement processes?

Answers on p. 261

Landforms and landscapes of coastal erosion

Cliffs and wave-cut platforms

The formation of cliffs and wave-cut platforms is outlined in Figure 3.7.

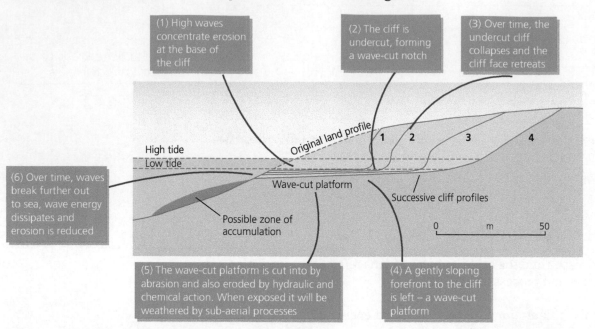

Figure 3.7 The formation of cliffs and wave-cut platforms

Headlands and bays

On any coastline there will be bands of rock of alternating resistance between more and less resistant rock. Two distinctive types of coastline develop:

+ A **discordant coastline** where bands of rock lie perpendicular (or at right angles) to the coastline. The weaker rock is eroded more rapidly to form a **bay, headlands** are left where the rock is more resistant, and erosion of the more-resistant headland is therefore slower.
+ A **concordant coastline** where bands of rock lie parallel to the coastline. A band of resistant rock on the coast will protect the weaker rock behind it. The coastline will therefore be quite straight.

Wave refraction (Figure 3.8) takes place as waves approach headlands and the energy is concentrated on the sides of the headland. This can lead to the formation of caves, arches and stacks.

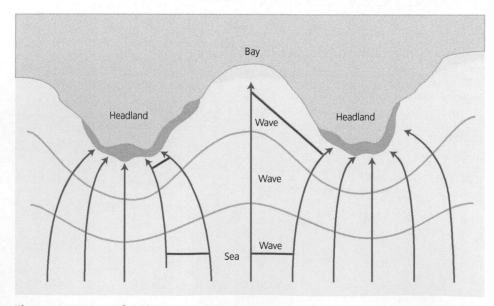

Figure 3.8 Wave refraction

Exam tip

Landforms are small-scale features; landscapes refer to the way features and landforms interconnect.

Wave-cut platform A gently sloping, smooth platform at the base of a cliff, caused by abrasion.

Now test yourself

21 Why is there a higher energy input on headlands?

22 Explain the difference between a concordant and a discordant coastline.

23 What factors affect coastal landscapes and their characteristic landforms?

Answers on p. 261

TESTED ○

Check your understanding and progress at **www.hoddereducation.co.uk/myrevisionnotesdownloads**

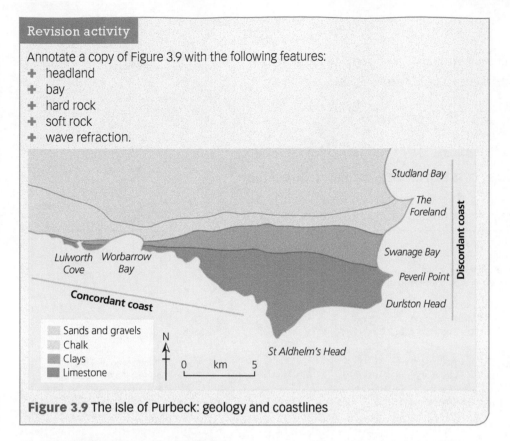

Figure 3.9 The Isle of Purbeck: geology and coastlines

Caves, arches and stacks

The formation of caves, arches and stacks is outlined in Figure 3.10.

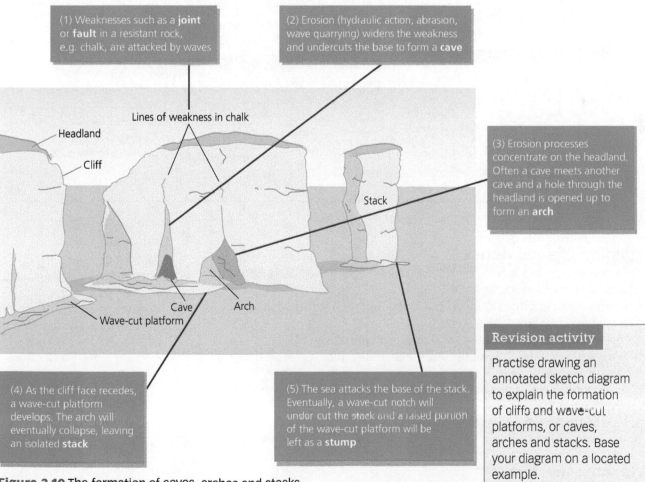

(1) Weaknesses such as a **joint** or **fault** in a resistant rock, e.g. chalk, are attacked by waves

(2) Erosion (hydraulic action, abrasion, wave quarrying) widens the weakness and undercuts the base to form a **cave**

(3) Erosion processes concentrate on the headland. Often a cave meets another cave and a hole through the headland is opened up to form an **arch**

(4) As the cliff face recedes, a wave-cut platform develops. The arch will eventually collapse, leaving an isolated **stack**

(5) The sea attacks the base of the stack. Eventually, a wave-cut notch will under cut the stack and a raised portion of the wave-cut platform will be left as a **stump**

Figure 3.10 The formation of caves, arches and stacks

Now test yourself   TESTED ◯

24 List the processes involved in the formation of caves, arches and stacks.
25 Describe the characteristics of a landscape of coastal erosion.

Answers on p. 261

Landforms and landscapes of coastal deposition

Beaches, spits, tombolos, offshore bars, barrier beaches and islands

The landscapes and landforms of coastal deposition are shown in Figure 3.11.

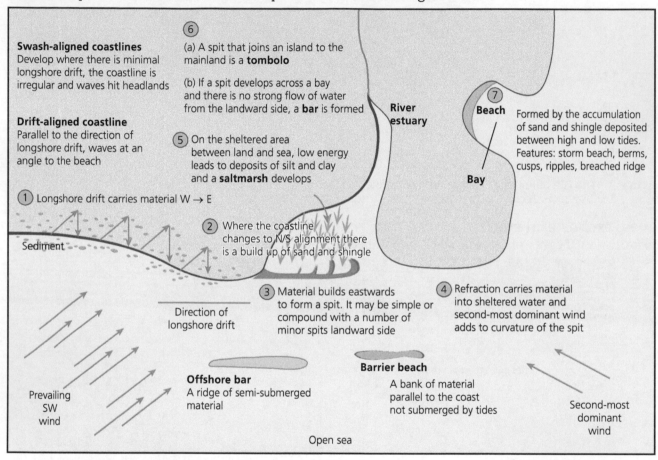

Figure 3.11 Landscapes and landforms of coastal deposition

Formation of sand dunes

Wind is the energy input into the formation of dunes – they are a **dynamic** landform. The main processes are:

+ **saltation** – a bouncing/skipping movement of small sand particles taking place up to 1 m above the surface
+ **creep** – the surface movement of larger sand particles.

Sand dunes require the following conditions to develop:

+ a plentiful supply of sand
+ a shallow offshore zone that allows large areas of sand to dry out at low tide
+ a wide backshore for sand to accumulate
+ prevailing onshore winds.

Sand dune morphology

+ When sand dries out on a beach, the wind blows it inland to form dunes (see Figure 3.12).
+ Initially, sand accumulates around pieces of detritus, such as litter or driftwood.

3 Coastal systems and landscapes

64

+ As the process of dune morphology takes place, vegetation such as marram grass anchors the dunes.

A typical sequence of sand dune development would be:

pioneer species → embryo dunes → foredunes (yellow) → fixed dunes (grey) → wasting dunes (slacks and blowouts)

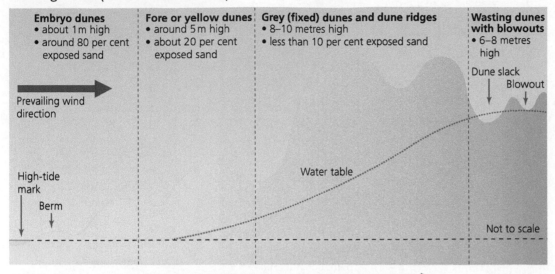

Age of dune increases with distance inland (≈400 years to maturity)

Figure 3.12 A typical sand dune transect

Now test yourself   TESTED ◯

26 Explain the process of longshore drift.

27 What is the difference between a swash-aligned beach and a drift-aligned beach?

28 Explain the sequential formation of spits, bars and tombolos.

29 What are the main processes in the formation of sand dunes?

Answers on p. 261

Exam tip

Sand dunes are an example of plant **succession**, the structure of which develops over time in stages. This is called a **psammosere**. It forms an ideal fieldwork study, which you can refer to in exam answers.

Key features of sand dunes are:
+ **ridges** – sequences of dunes parallel to the coastline
+ **slacks** – depressions which reach down to the water table and separate dunes
+ **blowouts** – depressions which form where fragile sand dunes have their vegetation cover destroyed by grazing animals such as rabbits or by human activity, for instance trampling.

Now test yourself   TESTED ◯

30 How does vegetation influence the formation of sand dunes?

31 What is a:
   a) dune slack
   b) blowout?

Answers on p. 261

Revision activity

Draw a cross section through a typical salt marsh and annotate it with features, processes at work and ecology. Figure 3.12 above shows a cross section – a slice through the landscape.

Landscapes of estuarine mudflats and salt marshes

Mudflats and **salt marshes** are landforms that form in sheltered low-energy coastlines (see Table 3.4). They are associated with large tidal ranges where powerful currents transport large quantities of fine sediment.

Table 3.4 Mudflat and salt marsh formation

Feature	Landscape	Processes	Ecology
Mudflats, e.g. Morecambe Bay, NW England	+ Low-lying, sheltered shorelines + Often an estuary or the landward side of a spit + Made of silt and clay. Flowing water forms permanent channels exposed at low tide + High salinity levels and low O_2 levels in mud	Flocculation – clay particles join and form heavier particles, which sink to the bed	Algae (no plants)
Salt marsh, e.g. Alnmouth, Northumberland	+ Flat, low-lying estuarine areas + Formed from mudflats over time	A process of vegetation succession called a halosere 1 Pioneer species develop (halophytes – salt tolerant) 2 Soil develops, lower salinity, current slowed, more deposition, organic matter produced. The marsh increases in height. Biodiversity and plant cover increase 3 Mud level rises, land rises above sea level, rushes and reeds grow. Salinity levels fall and soil develops	Glasswort (Salicornia), cord grass (Spartina). These plants slow the movement of water and encourage sedimentation. Their roots stabilise mud Sea aster, sea lavender, marsh grass Climax vegetation: ash, alder and oak

Sea-level changes

The average sea level relative to the land changes through time and this has happened in the last 10,000 years following the last glacial period 18,000 years ago.

Sea levels fall because of glacial periods and rise as climate warms. A typical sea-level change sequence is:

+ Stage 1: Sea levels fall due to storage of water as ice → **eustatic fall** in sea level.
+ Stage 2: The weight of the ice causes land surfaces to sink; some coastlines will have an **isostatic movement** – a moderation to the eustatic sea-level fall.
+ Stage 3: Climate warms, ice melts, sea levels rise globally → **eustatic change**.
+ Stage 4: Ice is removed from some land areas which move back to their original levels before the weight of the ice → **isostatic adjustment**.

Coastlines of emergence and submergence

Key terms are:

+ **Eustatic change** – global change in sea level resulting from a rise or fall in the level of the sea itself, for example due to the retreat of ice following a glacial period.
+ **Isostatic change** – local change in sea level resulting from the land rising or falling relative to the sea, for example tectonic movements.

Dalmatian coasts

These are similar to rias except that the rivers flow more parallel to the coast, whereas with rias the river flow is at right angles to the coast. An example is Croatia.

Raised beaches, marine platforms, rias and fjords

Figure 3.13 shows the features formed by a rising and falling sea level.

Check your understanding and progress at **www.hoddereducation.co.uk/myrevisionnotesdownloads**

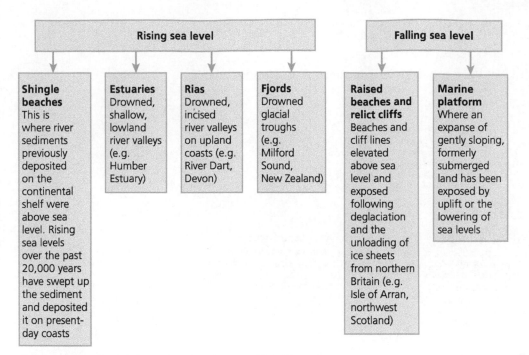

Figure 3.13 Changing sea level and coastal landforms

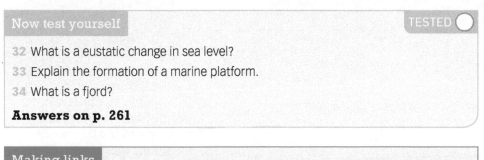

Making links

See Chapter 1 page 28 for links between sea-level change and climate change.

Revision activity

From the examples used in class, draw an annotated sketch map of the landscape and landform features of:

a) an emergent coastline, for instance the Isle of Arran

b) a submergent coastline, such as Croatia.

Recent and predicted climate change and the potential impact on coasts

Two processes are causing current sea-level changes:

✚ an increase in the volume of oceans due to melting ice caps and thermal expansion

✚ subsidence of the coast.

Predicting climate change is difficult as it involves complex modelling. However, potential impacts of rising sea levels on coasts could include:

✚ increased coastal flooding, particularly with spring tides and strong onshore winds

✚ more coastal erosion and cliff erosion as waves attack areas previously above high tide

✚ receding coastlines

✚ an upstream movement of the zone where seawater mixes with fresh water in rivers

✚ increased flooding of areas away from the coastline as rivers flood more

✚ greater frequency and magnitude of extreme sea-level events, such as storm surges

✚ increased erosion of dunes, salt marshes and mudflats

✚ increased investment in coastal protection for areas of high economic value.

The relationship between process, time, landforms and landscapes in a coastal setting

Figure 3.14 outlines the link between process, time, landforms and landscapes in a coastal setting.

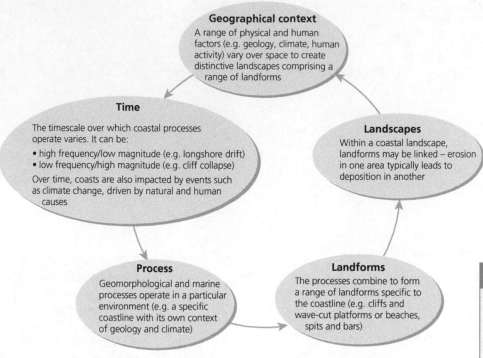

Geographical context
A range of physical and human factors (e.g. geology, climate, human activity) vary over space to create distinctive landscapes comprising a range of landforms

Time
The timescale over which coastal processes operate varies. It can be:
• high frequency/low magnitude (e.g. longshore drift)
• low frequency/high magnitude (e.g. cliff collapse)
Over time, coasts are also impacted by events such as climate change, driven by natural and human causes

Landscapes
Within a coastal landscape, landforms may be linked – erosion in one area typically leads to deposition in another

Process
Geomorphological and marine processes operate in a particular environment (e.g. a specific coastline with its own context of geology and climate)

Landforms
The processes combine to form a range of landforms specific to the coastline (e.g. cliffs and wave-cut platforms or beaches, spits and bars)

Figure 3.14 Coastal setting: the link between process, time, landforms and landscapes

Now test yourself

TESTED ⬤

35 What will be the impact of rising sea levels on coastlines?

36 What are the economic impacts of a receding coastline?

37 Give an example of a high-frequency, low-magnitude coastal process.

Answers on p. 261

Revision activity

Either from your own notes or from research, produce a summary sketch of a located stretch of coastline, label the coastal landforms that make up this landscape and also add information on the distinctive climatic, geological and human factors that contribute to the landscape development. Examples include Flamborough Head, Robin Hood Bay and Chesil Beach.

Coastal management

REVISED ⬤

Human intervention in coastal landscapes

Coastal management is a response to the natural and human activities that threaten coastal environments, for example erosion, flooding, overdevelopment and pollution. The aims of coastal management are outlined in Figure 3.15.

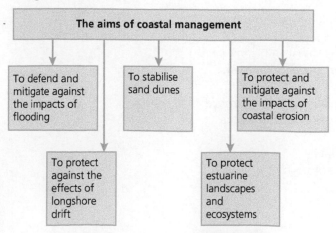

The aims of coastal management

To defend and mitigate against the impacts of flooding

To stabilise sand dunes

To protect and mitigate against the impacts of coastal erosion

To protect against the effects of longshore drift

To protect estuarine landscapes and ecosystems

Figure 3.15 The aims of coastal management

Check your understanding and progress at **www.hoddereducation.co.uk/myrevisionnotesdownloads**

Traditional approaches to coastal flood and erosion risk: hard and soft engineering

Hard engineering

Hard engineering involves the construction of a variety of structures (see Table 3.5). It can be expensive and it can have negative impacts further along the coastline.

Table 3.5 Hard engineering coastal defences

Sea walls	Concrete Beach material Steel pile	Provide a physical barrier to flooding and recurved walls dissipate wave energy
Revetments	Open structure of planks to absorb wave energy but allow water and sediment to build up beyond	Concrete or wooden structures placed along a coastline to absorb wave energy
Groynes	Wood or steel piling Concrete wall Beach material	Wooden, stone or steel breakwaters placed just off right angles to waves to trap sediment and control longshore drift
Rip-raps	Large boulders positioned on beach	Concrete blocks or boulders placed at the foot of cliffs to take the wave impact
Gabions	Steel wire-mesh cage filled with small rocks	Small boulders contained in a wire-mesh cage to take wave impact
Barrages	Pier Lifting mechanism Hydraulic system Gate River bed	Large structures built to prevent flooding on major estuaries

> **Exam tip**
>
> Any intervention in the coastal system invariably has an impact elsewhere, for example by accelerating erosion or narrowing beaches.

Soft engineering

Soft engineering seeks to work with the natural coastal processes. It is cheaper than hard engineering and can be more environmentally sustainable (see Table 3.6).

Table 3.6 Soft engineering coastal defences

Beach replenishment	Replacement of sediment/sand lost through longshore drift
Managed retreat	Abandoning of current sea defences and management of exposed land to reduce wave power, e.g. salt marshes, mudflats
Dune regeneration	Stabilisation of dunes by planting of marram grass, selective grazing, providing boardwalks for tourists, brushwood barriers to encourage sand accumulation
Marsh creation	Realignment of the coast, e.g. by creating salt marshes to absorb wave energy

Revision activity

From the located examples you have studied in class, complete two tables to summarise the two approaches to coastal management:

+ One table for hard engineering, for example Heysham, Morecambe, Lancashire or North Bay, Scarborough.
+ One table for soft engineering, for example Formby Point, Sefton coast, Lancashire or Freiston Shore, Lincolnshire.

Include columns for technique, description and evaluation.

Sustainable approaches to coastal flood and erosion risk

Shoreline management plans

The strategy of **shoreline management plans (SMPs)** uses a combination of approaches that aim to be cost effective and long lasting.

+ **Hard engineering** is used in areas where erosion or flooding could cause greatest economic loss.
+ **Soft engineering** is used where there may be a fragile ecosystem or where it is possible to work with natural processes.
+ The overall management plan aims to be economically, socially and environmentally sustainable.

SMPs comprise three strategies:

+ **Hold the line** – existing defences are maintained (or strengthened).
+ **No active intervention** – natural processes are allowed to operate without human intervention.
+ **Managed realignment** – coastal areas that were previously protected or had been reclaimed are deliberately flooded. Salt marshes and mudflats can then naturally trap sediment and create natural defences with new habitats for wildlife, for example Blackwater Estuary, Kent.

The basic units for SMPs in England and Wales are within the 11 sediment cells shown in Figure 3.5. Within these, lengths of coast known as sub-cells are identified for management.

Integrated coastal zone management plans

Integrated coastal zone management (ICZM) plans originated from the UN Earth Summit, Rio in 1992. The aims are set out in the Agenda 21 documents. The European Commission states that the aim of ICZM is to:

> contribute to the sustainable development of coastal zones by the application of an approach that respects the limits of natural resources and ecosystems, the so-called 'ecosystem-based approach'.

ICZM is designed to integrate the views and interests of all **stakeholders** in a management issue and to co-ordinate policies that affect the coastal zone, such as fishing, agriculture, industry and offshore energy.

Exam tip

You need to revise carefully the examples of approaches to coastal management in this section, so that you can give a thorough evaluation in a high-level response.

Exam tip

It is important to know the name of an example of an ICZM plan, and some of the advantages and disadvantages of this approach, which seeks to involve all interested stakeholders.

Check your understanding and progress at **www.hoddereducation.co.uk/myrevisionnotesdownloads**

Identify a length of coastline where there is an SMP in place – use Figure 3.5, a map of sediment cells around England and Wales. On A3 paper, produce an annotated map of the plan. Include bullet point information on each technique and colour-code the different phases/features of the plan. An example is Sea Palling on the Norfolk coast.

Make summary notes of your local coastal case study. This should show fundamental coast processes and engage with field data.

In addition to a sketch map, use the following headings for your notes:

✚ Coastal processes – what are the main geomorphological and specific marine processes at work on this stretch of coastline? Factors such as climate, prevailing wind direction and geology will be important considerations.

✚ Landscape outcomes – a summary of the characteristic landscape and specific landforms.

✚ Challenges – what are the present and future physical and human challenges on this coastal stretch?

✚ Sustainable management – what are the opportunities for sustainable management? How are they being met presently/in the future?

Case studies

A coastal environment at a local scale, e.g. Pevensey Bay, East Sussex

This case study should:

✚ illustrate and analyse fundamental coastal processes and their landscape outcomes
✚ engage with field data
✚ address the challenges represented in sustainable management.

What do I need to know?	Content and suggested revision methods

Coastal processes active in the chosen location and their landscape outcomes	You can organise your revision notes for this section into a detailed, **annotated map**. This may be a sketch map, or a suitable base map used in fieldwork.
	
	Annotations can:
	✚ explain the coastal processes active in the area:
	  ✚ prevailing wind direction and its impact on the process of longshore drift or on the intensity of erosion
	  ✚ how different erosion processes impact on the coast (this can be related to the geology of the area)
	✚ describe the resulting landscape features:
	  ✚ beaches
	  ✚ storm beaches
	  ✚ shingle banks.
	
	In order to cover detail, you may need to use A3 paper.
	Annotations may be supported with fieldwork data.

A range of strategies for sustainable beach management	It is best to organise revision notes into a table. This should name the strategy for coastal defence, outline how it works and state how it is sustainable. For example:

Strategy	Description	How is this sustainable?
Recharge	20,000 m³ of gravel and sand have been left in situ offshore to be sorted and redistributed by longshore drift.	Instead of using noisy, polluting and costly bulldozers to push the sand and gravel into place, there is natural sorting which saves time and money.

The Pevensey Bay sea defences project aims to manage the coastline sustainably by focusing on a **soft management approach**, working with coastal processes.

A contrasting coastal landscape beyond the UK, for example the Sundarbans, Bangladesh

This case study should:
+ illustrate and analyse how it presents risks and opportunities for human occupation and development
+ evaluate human responses of resilience, mitigation and adaptation.

What do I need to know?	Content and suggested revision methods
The risks/challenges and opportunities presented by living in the Sundarbans	Summarise your revision notes in a suitable diagram or table. The example below shows how colour coding subcategories can aid revision. + Opportunities are subdivided into goods (red) and services (blue). + Challenges are subdivided into natural challenges (brown) and human challenges (purple). **Opportunities** • Fuel • Construction materials • Fishing materials • Household items • Food and drink • Textiles • Other products • Protection • Provision • Maintenance • Value **Natural challenges** • Coastal flooding • Cyclones • Highly saline soils • Instability of the islands • Remoteness • Human-eating tigers **Human challenges** • Over-exploitation of coastal resources • Intensification of agriculture • Destructive fishing techniques • Resource-use conflicts • Lack of awareness of coastal issues and the economic and environmental importance of the area
1 Human resilience to living in the Sundarbans 2 Attempts to mitigate/moderate the risks of living in the Sundarbans 3 Human adaptation to life in the Sundarbans	There are three clear sections to this part of the case study. You can either choose a different method to organise revision notes for each subsection, e.g. table, spider diagram, bullet points, or combine all three into one large diagram/mind map, e.g.: 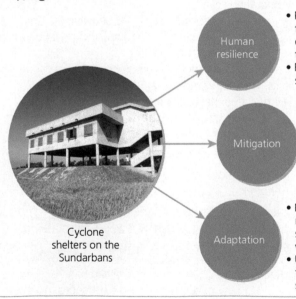 Cyclone shelters on the Sundarbans **Human resilience** • Protection by mangrove forests against storms, erosion on the coast, tsunamis, floods • Fertile soils provide a good supply of nutritious food **Mitigation** • Investment in physical infrastructure, e.g. cyclone shelters, flood protection • High level of community action **Adaptation** • Fertiliser usage can be reduced by growing new, salt-tolerant rice varieties which survive in sea water • Use of more storage tanks to adapt to changes to seasonal pattern of rainfall

Exam practice

1 Outline the role of wind energy in coastal systems. [4]

2 Explain how sediment budgets can affect coastline development. [4]

3 Identify and explain the relative importance of different processes in the formation of one of the coastal landforms in Figure 3.16. [6]

Figure 3.16 Coastline at Flamborough Head

4 Compare and contrast the features and effectiveness of hard and soft engineering coastal defences. [6]

5 Explain the role of vegetation in the development of salt marshes. [6]

Answers and quick quizzes online

Exam skills

Opportunities to practise geographical skills within this topic include:
+ observation skills in fieldwork
+ collection, manipulation, presentation and analysis of primary and secondary fieldwork data – including quantitative and qualitative sources (particularly in relation to the first case study)
+ statistical tests applied to data (for example primary and secondary sources in fieldwork)

+ analysis of specific graphs and histograms showing, for example, climatic data, sediment budgets
+ interpretation of geospatial data (for example geological maps)
+ simple data manipulation (for example calculation of sediment budgets).

Summary

+ The coastal system is an open, dynamic system driven by the energy of wind, waves, currents and tides.
+ Geomorphological, marine and sub-aerial processes lead to the formation of characteristic coastal landforms, which combine and integrate to form coastal landscapes unique to location and geographical contexts.
+ The sources and flows of sediment are key subsystems.

+ Rising sea levels impact human activities and coastal landforms.
+ Coastal erosion and flooding require management. Decisions regarding appropriate and sustainable management are complex and require careful evaluation.

4 Glacial systems and landscapes

Glaciers as natural systems

Systems in physical geography

The **systems** approach is a way of analysing the relationships within a unit, for example a glacier. It consists of a number of components and the linkages between them represented in a flow diagram.

A glacier is an **open system** (see Figure 4.1). It has:

+ **inputs** (kinetic energy from wind and from moving glaciers, thermal energy from the Sun)
+ **stores/components** (accumulated debris from erosion, deposition and weathering and ice), **flows/transfers** and
+ **outputs** of both energy and matter (erosion of rock surfaces, evaporation, meltwater), which cross the boundary of the system to the surrounding environment.

> **Glacier** A large mass of ice moving downhill under the influence of gravity.

The combination of all these factors forms distinctive landscapes, which are made up of a range of erosion and depositional landforms.

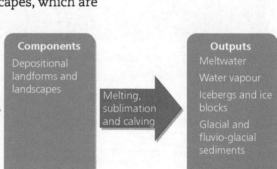

Figure 4.1 A glacier is an open system within a glacial landscape

Glaciers are dynamic (constantly changing):

+ The system is in a state of **dynamic equilibrium** with a balance between inputs and outputs.
+ Change occurs to upset the balance of the system – for a glacier this may be due to snowfall (accumulation) or melting/sublimation (ablation), for example.
+ The system adjusts by a process of **feedback**, which can be either:
  + **positive** (progressively greater change from the original condition of the system) or
  + **negative** (the system is returned to its original condition).

The glacial budget

The **glacial budget (mass balance)** is the difference between annual accumulation and annual ablation (see page 79). There are three states of mass balance in glaciers:

+ **Positive** – accumulation exceeds ablation and there is an increase in ice mass, causing the glacier to advance.
+ **Negative** – ablation exceeds accumulation, causing the glacier to retreat.
+ **Neutral** – accumulation and ablation are equal, so the glacier is in a static state.

Positive feedback is associated with the upper part of the glacier, referred to as the **zone of accumulation**.

Check your understanding and progress at **www.hoddereducation.co.uk/myrevisionnotesdownloads**

Negative feedback is associated with the lower part of the glacier – **the zone of ablation**.

The boundary between the two is the **firn line**.

Now test yourself TESTED

1 What are the sources of energy for a glacial system?
2 Explain the concept of dynamic equilibrium in a glacier.
3 What is a glacial budget?

Answers on p. 261

Revision activity

Produce a flow diagram to explain a positive and negative feedback in a glacial system.

The nature and distribution of cold environments

REVISED

The global distribution of cold environments

Cold climates supporting glaciers, ice sheets and permafrost occur in areas of high latitude and high altitude. The largest expanses of snow are in the polar regions, such as Antarctica and Greenland. Mountain ranges with glaciers include the Himalayas. Periglacial areas cover areas such as Alaska.

Physical characteristics of cold environments

Climate

Features of **polar** climates:

+ Mean monthly temperature below freezing all year.
+ Winter average < –50°C.
+ Precipitation 150 mmy^{-1} (snowfall); at the South Pole it can be just 50 mmy^{-1}.
+ Strong winds blowing outwards from the centre of the continent.

Features of **tundra** climates:

+ Winter average –20°C.
+ Brief summer of +5°C.
+ Eight months (at least) below 0°C.
+ <300 mmy^{-1} precipitation (snowfall).
+ Significant wind chill.

Exam tip

'Cold environments' is a broad term covering a range of locations. Be aware of this in responding to exam questions.

Causes of such climates

+ Low levels of insolation – low angle of the Sun, meaning there is a wide surface area to heat. Long, dark winters when there is no incoming solar radiation.
+ High albedo – reflection of heat by the snow cover.
+ High-pressure systems dominate; there are few frontal systems and low levels of precipitation.
+ Very cold air, which does not hold as much water vapour as warm air.
+ Katabatic winds – masses of cold, dense air flow down valleys.

Making links

See Chapter 1 page 28 to develop understanding on how ice-covered surfaces affect climate change and the role of feedback.

Soils

Tundra soils:

+ Lack of clear layers (horizons); organisms act as mixing agents, but it is also too cold for this activity.
+ Thin surface layer, which is very acidic.
+ Blue/grey colour due to grey ferrous iron compounds.
+ Waterlogged soils in summer.
+ Little activity from soil organisms due to the cold.

Exam tip

Remember that the features of climate, soil and vegetation are interconnected – climate impacts on vegetation and soil development, for example. Do not view these characteristics in isolation.

Now test yourself

TESTED

4   Name three differences between polar and tundra climates.
5   What is thermal insolation?
6   What is albedo?

Answers on p. 261

Vegetation

Tundra vegetation:

+ Very low productivity and slow growth rates.
+ Low biological diversity, leading to simple and therefore fragile food chains and webs.
+ Absence of full-grown trees.
+ Only perennial flowering plants; perennial plants can store food from year to year.
+ Plants comprise mainly lichens, mosses, grasses, cushion plants and low shrubs.
+ Low-height vegetation, to avoid strong winds.
+ Shallow roots to capture water from the spring thaw.
+ Ability to carry out photosynthesis at very low temperatures.
+ Thick cuticles and small leaves to cut down transpiration.

The global distribution of past and present cold environments

Cold environments refer to a range of landscapes, climates and ecosystems. There is also variation in the meaning of the term 'cold'. Cold environments can be grouped into four types – polar, alpine, glacial and periglacial.

Polar

Polar climates are associated with the northern and southern extremes of the Arctic and Antarctic (Table 4.1).

Table 4.1 The characteristics of polar climates

	Characteristics
Arctic	+ Circles the North Pole and the northerly extremes of Asia, Europe and North America + Mean temperature range of –28°C to 4°C + Average annual precipitation of around only 100 mm
Antarctic	+ Much colder than the Arctic due to strong westerly winds, cold oceans and a large landmass + Mean temperature of –55°C in places. Coastal areas of Antarctica are milder with an annual average of –10°C + An average annual precipitation of 200 mm + Winter sea ice around Antarctica is increasing while in the Arctic it is shrinking

Alpine

+ **Relief:** refers to areas of high relief, generally over 3000 m, for example Himalayan and Tibetan mountain ranges in Asia, the Rockies and the Andes in the Americas, and the New Zealand Alps.
+ **Temperature:** temperatures range from –10°C in winter to 20°C in the summer months.
+ **Landscapes:**
   + include ice caps, mountain glaciers and tundra
   + develop over glacial and interglacial periods. A combination of tectonic uplift and rapid rates of erosion by water and ice creates well-developed glacial landforms.
+ **Geographical location:** alpine regions with native glaciers occur today at any latitude where altitude is high enough for snow and ice to remain throughout the year.

Check your understanding and progress at **www.hoddereducation.co.uk/myrevisionnotesdownloads**

Glacial

These are areas currently covered by ice sheets and glaciers. There is ice throughout the year. They include the Antarctic and Greenland.

Periglacial

This category includes tundra regions.

Periglacial regions are areas of dry, high latitudes, which may not be permanently covered by snow and ice but are areas of permafrost overlain by an 'active' layer of soil.

Areas include North Alaska and Canada, northern Scandinavia and Siberia (see Figure 4.2).

> **Revision activity**
>
> Create a spider diagram showing the features of and links between climate, soils and vegetation in cold environments.

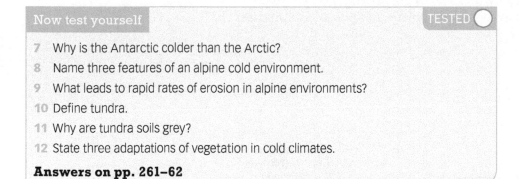

> **Exam tip**
>
> If a question is based on a data source for a cold environment area that you are unfamiliar with, you just need to apply your knowledge and understanding of the topic. For example, 'Using Figure xxx and your own knowledge... ' requires you to apply your knowledge and understanding to the data and doesn't expect you to have knowledge about the specific example that the data relates to.

> **Now test yourself**   TESTED ○
>
> 7   Why is the Antarctic colder than the Arctic?
> 8   Name three features of an alpine cold environment.
> 9   What leads to rapid rates of erosion in alpine environments?
> 10 Define tundra.
> 11 Why are tundra soils grey?
> 12 State three adaptations of vegetation in cold climates.
>
> **Answers on pp. 261–62**

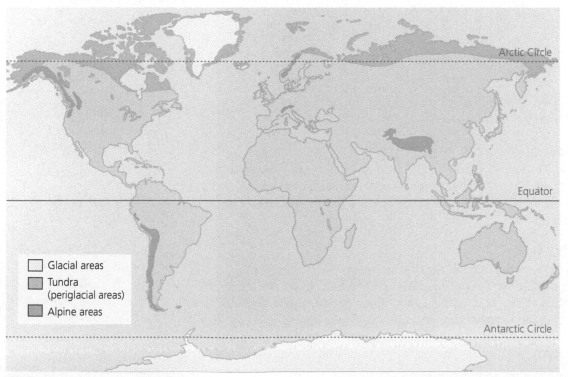

Figure 4.2 Cold environments

Areas affected by the Pleistocene glaciations

The Pleistocene glaciations cover the period between 1.6 million and approximately 12,000 years BP (before present). In this time glacial and interglacial periods have resulted in cold environments expanding and contracting.

> **Revision activity**
>
> You should know the distribution of cold environments. Annotate a copy of Figure 4.2 with brief bullet point notes on the location and characteristics of polar, alpine, glacial and periglacial environments.

Systems and processes

Glacial systems and glacial budgets

Glacial systems

Glaciers are open systems. Inputs include direct snowfall, blown snow and avalanches. Together, these inputs are known as accumulation.

The inputs are transferred (down valley) by gravity. Mass is lost from the system by melting and evaporation – ablation. Figure 4.3 summarises the glacial system.

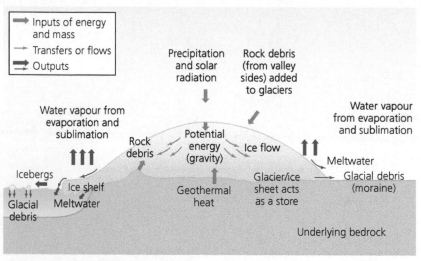

Figure 4.3 The glacial system

The formation of glaciers

Snow initially falls as flakes. On settling, the lower layers compact to **firn** or névé (French term for firn). With further compaction, air is forced out and a bluish colour is evident. A large mass of ice forms a glacier and starts to move downhill due to gravity.

The glacial budget

As a glacier is an open system, the mass will increase and decrease depending on the balance of accumulation and ablation. In winter there is more accumulation than ablation. The **snow line** represents the boundary between snow-covered areas and areas where a higher temperature means no snow cover. The snow line is a seasonal adjustment moving down slope as temperatures drop in winter.

The glacial budget is depicted in Figure 4.4.
+ The **zone of accumulation** => upper part of a glacier where input > output.
+ The **zone of ablation** => where output > input.
+ The boundary is the **equilibrium line**.

> **Making links**
>
> See Chapter 1 page 13 for more on cryospheric processes.

> **Revision activity**
>
> Using Figure 4.3 and your own notes, practise answering the following question: 'Outline the different elements that make up a glacial system.'
>
> You may present your answer as a diagram, flow chart or written text. Remember to have a clear structure and sequence to your answer.

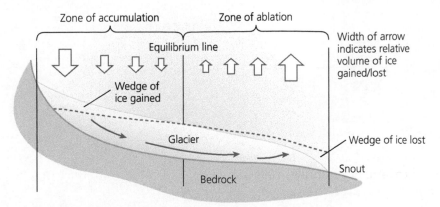

Figure 4.4 Glacial budget

Ablation and accumulation – historical patterns of advance and retreat

The difference between accumulation and ablation in a year is the **net balance**:

+ In times of high levels of accumulation, glacial advance occurs.
+ In times of high levels of ablation, glacial retreat occurs.

During the glacial and interglacial periods of the Quaternary, glaciers have advanced and retreated many times. At present glaciers cover 6500km² in Europe; 18,000 years BP, ice coverage was much greater.

Individual glaciers respond to long-term climate change but also to short-term changes. Most have retreated in the recent past, but there are some exceptions – the Hubbard Glacier in Alaska, for example, has advanced.

Warm- and cold-based glaciers

Glaciers can be classified as warm or cold based.

Warm-based glaciers

Warm-based glaciers occur in temperate areas, such as western Norway and southern Iceland.

+ They are small (hundreds of metres to a few kilometres in width).
+ There is a summer melt.
+ Meltwater lubricates the glacier, leading to more movement and consequently more erosion, transportation and deposition.
+ All ice in warm-based glaciers is at or near melting point due to the warmer atmospheric temperature, the weight of ice and the effect of geothermal heat at the bed.
+ Basal temperatures are at or above pressure melting point.

Cold-based glaciers

Cold-based glaciers occur in polar areas, such as the Arctic and Antarctic.

+ They are large, vast ice caps and ice sheets covering hundreds of km².
+ They occur in areas of low precipitation and little snow. Therefore, there are low levels of accumulation and no melting as the ice stays very cold.
+ All ice is below melting point.
+ There is very little meltwater and therefore slow movement. The glacier is often frozen to the bed of the glacier, meaning less erosion, transport and deposition.
+ Basal temperatures are below pressure melting point.

Now test yourself

13 What is the snow line?
14 What is the equilibrium line?
15 What is firn?
16 Why is there more erosion with a warm-based glacier?

Answers on p. 262

TESTED

Pressure melting point
The temperature at which ice melts under pressure.

Revision activity

Make a small pack of cards. On each card write a feature of either warm- or cold-based glaciers. Mix the cards and then sort them into two piles – one for features of warm-based glaciers and the other for features of cold-based glaciers.

17 Explain the term 'net balance'.

18 How will 'net balance' vary over time?

Answers on p. 262

Geomorphological processes

Weathering

Table 4.2 summarises the physical, chemical and biological weathering processes operating in cold environments.

Table 4.2 Physical, chemical and biological weathering processes operating in cold environments

Process	Description
Physical/mechanical weathering	
Frost shattering	+ Water trapped in tiny pores in rocks freezes in very cold environments. + As it freezes, it expands and small pieces of rock become dislodged/shatter.
Freeze–thaw	+ Water enters cracks in rocks during the day and freezes overnight. + As it freezes, it expands by approximately 10%, exerting pressure. + As this process is repeated, it widens the crack and leads to rock fragments breaking off.
Pressure release	+ During melting, the weight of ice is reduced. + This reduction in pressure causes the rock to expand and fracture or break up near the surface. + This also allows meltwater in for freeze–thaw action.
Chemical weathering	
Hydration	+ The breakdown of rock by cycles of expansion during wetting and contraction during drying. + When water molecules and rock minerals are joined together, they form minerals of a larger volume.
Oxidation	+ Some minerals in rock react with oxygen that is absorbed from either air or water. + A chemical reaction takes place that weakens certain types of rock (e.g. iron elements of sandstone).
Carbonation	+ Rainwater combines with dissolved CO_2 from the atmosphere to produce a very weak carbonic acid. + This reacts with calcium carbonate in rocks, such as limestone, to produce calcium bicarbonate, which is soluble. + This leads to chemical breakdown of rock.
Solution	+ The process whereby minerals in rocks (e.g. iron) dissolve in water, thereby weakening the rock structure.
Biological weathering	
Tree root expansion	+ Tree roots grow into joints and cracks in rocks and exert pressure. Over time, this will loosen rock particles. + If trees fall in strong winds, the roots will exert extra pressure.
Organic acids	+ Decomposition of plant and animal matter produces organic and soil water becomes more acidic and reacts with some rock minerals (e.g. iron).

19 What is the difference between physical and chemical weathering?

Answer on p. 262

Check your understanding and progress at **www.hoddereducation.co.uk/myrevisionnotesdownloads**

Ice movement

Revision activity

Make a summary spider diagram of the geomorphological processes associated with cold environments. See page 60 for the one on coastal processes. Use the following as bubble headings:
+ Weathering
+ Erosion
+ Transportation
+ Deposition
+ Mass movement

Ice moves under the force of gravity. There are two zones of movement:
+ **upper zone:** brittle, breaking and forming crevasses
+ **lower zone:** steady pressure, meltwater from pressure melting and frictional heat leads to more rapid movement.

Table 4.3 details the types of ice movement.

Table 4.3 Types of ice movement

Type of movement	Description
Internal deformation	Ice crystals orientate themselves in the direction of the glacier movement and slide past each other. This is the main type of flow in cold glaciers in the absence of meltwater.
Rotational flow	This occurs within the corrie (cirque) or depression in which the glacier forms – the ice rotates/pivots as it starts to move downhill.
Compressional flow	Reduction in gradient leads to a thickening of ice mass and slowing of movement (see Figure 4.5).
Extensional flow	A steeper gradient leads to the thinning and acceleration of the ice (see Figure 4.5).
Basal sliding	As the ice moves over bedrock there is friction, pressure and therefore heat. The heat leads to melting. The resulting meltwater acts as a lubricant and the ice flows more rapidly.

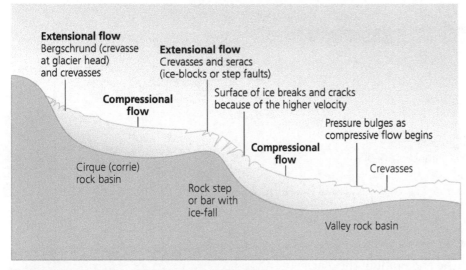

Figure 4.5 Extensional and compressional flow

Erosion

Abrasion

Abrasion occurs when material in the glacier rubs away at the valley sides and floor. Scratches may be left – these are known as striations. If the debris is very fine, it is called rock flour.

Plucking

Plucking occurs when the glacier freezes onto and into rock outcrops. As the ice moves, it pulls away pieces of rock. This mainly occurs at the base of the glacier where jointed rocks have been weakened by freeze–thaw action. Plucking leaves a jagged landscape.

Nivation

Nivation is neither classified as weathering nor as erosion; it is a series of processes, including physical and chemical weathering, that operate underneath patches of snow. Freeze–thaw action and chemical weathering processes loosen rock and meltwater removes the debris. This repeated melting and freezing and removal of debris over seasons forms **nivation hollows**.

> **Revision activity**
>
> Rates of abrasion in cold environments are affected by a range of factors. For each of the factors listed below, explain how they would affect abrasion:
> + amount of debris at the base of a glacier
> + debris size and shape
> + ice thickness
> + sliding of basal ice.
>
> For example:
>
> | Amount of debris at the base of a glacier → | Abrasion would increase as there is more material to be carried by meltwater and used in the process of abrasion. |
> | --- | --- |
>
> You can do this activity as a simple table as shown above or as a flow or line diagram.

Transportation

Glaciers carry large amounts of debris from weathering and erosion processes and rockfalls from the valley side. Transportation is:
+ **supraglacial**: on the surface
+ **englacial**: within the ice
+ **subglacial**: at the base of the glacier.

> **Now test yourself** TESTED
>
> 20 How does ice move in:
>   a) the upper zone
>   b) the lower zone?
> 21 Define extensional flow and compressional flow.
> 22 Name the three ways in which glaciers transport debris.
>
> **Answers on p. 262**

Deposition

Material is deposited:
+ when the ice melts at the **snout** (the end point of the glacier)
+ where the glacier changes between compressing and extending flow.

Glacial till

Till is a term used to describe unsorted rocks, clay and sand debris. Till is angular, unsorted and unstratified (that is, dropped in mounds rather than layers).

The composition of till reflects the geological conditions over which the ice has travelled.

Fluvioglacial processes

Meltwater is formed when glaciers melt. Large quantities of meltwater are produced, which transport large amounts of debris.

Meltwater channels are typically steep sided, deep and straight. With high discharge and turbulent flow, large meltwater channels have significant levels of erosive power.

Periglacial features and processes

Periglacial areas are not glaciated but are exposed to very cold conditions with intense frost and permanently frozen ground.

Permafrost

Permafrost is the permanently frozen ground in areas where temperatures below the ground surface remain below 0°C continuously for more than two years. There are three categories of permafrost:

+ **continuous** (in the coldest regions of Siberia it can be 1500 m thick)
+ **discontinuous** (in slightly warmer regions it is 20–30 m thick)
+ **sporadic** (isolated spots of permafrost occur in areas where there is a summer thawing).

In the summer, temperatures rise above freezing and there is a brief summer thaw of the surface layer, which becomes known as the **active layer**.

Mass movement

Mass movement processes that occur in periglacial areas include:

+ **solifluction** – in the summer, water in the surface layer melts but, due to an impermeable frozen layer below it, cannot drain away; also lack of evaporation means that the surface layer becomes very wet, soil particles become lubricated and will move down the most gentle of slopes
+ **frost creep** – the gradual downslope movement of individual soil particles due to freeze–thaw cycles
+ **rockfalls** – the movement of large amounts of scree produced by freeze–thaw weathering
+ **slides or slumps** – common in weaker rocks (for example clay, which becomes heavy when wet).

Now test yourself

23 What are the key features of glacial till?

24 Why is solifluction more active in summer months?

Answers on p. 262

TESTED

Glaciated landscape development

REVISED

Erosional landforms

Corries and arêtes

Corries (cirques) are armchair-shaped rock basins with a rock lip that are cut into mountains (see Figure 4.6). Mostly they occur on north- and east-facing slopes where less insolation allows snow to accumulate. Their formation can be sequenced as follows:

1 Freeze–thaw weathering above the glacier creates debris, which falls onto the top of the glacier.
2 Abrasion occurs at the base of the glacier as it flows forwards and downslope. The depression is deepened.
3 Plucking steepens the back wall and adds debris. The back wall retreats further due to freeze–thaw weathering.
4 Meltwater flows down a deep crevasse (bergschrund), which opens up between the glacier and the back wall. This water aids movement (as it lubricates the base of the glacier). Abrasion continues and as the meltwater freezes, plucking and freeze–thaw processes take place also.
5 The glacier moves in a rotational pattern and continued erosion deepens the basin.
6 At the outlet of the basin, ice movement is upwards, there is less erosion and a lip forms.

Exam tip

In your explanations, always follow a clear sequence of events and refer to named processes rather than just saying 'by erosion'. Make clear the link between processes and landforms.

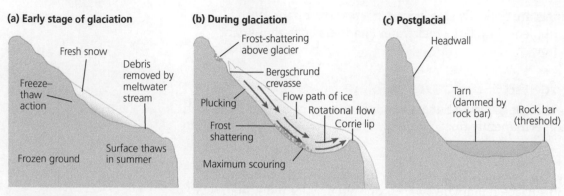

Figure 4.6 The formation of a corrie

An **arête** is formed where two corries lie back to back. If more than two corries develop on a mountain, the remaining central mass forms a **pyramidal peak**.

Examples of the above landforms include:

+ **corrie:** Red Tarn Corrie, Lake District
+ **arête:** Striding Edge above Red Tarn, Lake District
+ **pyramidal peak:** Machhapuchhre, Nepal.

Glacial troughs, hanging valleys, truncated spurs and roche moutonnées

Glaciers flow down pre-existing river valleys and, because of their power, they deepen these valleys and change the V to a U shape (see Figure 4.7). These U-shaped valleys are straight, wide based and steep sided. They are known as **glacial troughs**.

The ice, meltwater and subglacial debris combined have huge erosive power. Several further landforms are associated with glacial troughs:

+ **Truncated spurs:** the glacier removes areas of land protruding from the river valley side (spurs), forming truncated spurs.
+ **Roches moutonnées:** the glacier doesn't completely remove areas of resistant rock on the valley floor, which are left as roches moutonnées. They have a smooth up-valley side created by abrasion and a jagged down-valley side due to the action of plucking.
+ **Ribbon lakes:** After the ice retreats, ribbon lakes are formed in rock basins in the base of a glacial trough.

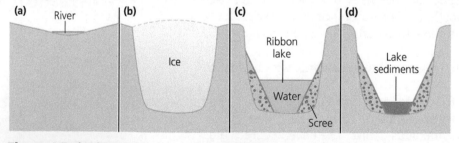

Figure 4.7 The formation of a glaciated valley

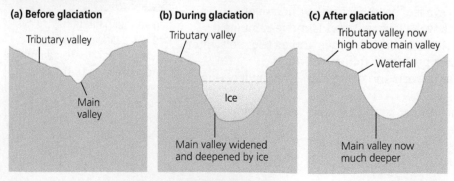

Figure 4.8 The formation of a hanging valley

Check your understanding and progress at **www.hoddereducation.co.uk/myrevisionnotesdownloads**

Examples of these landforms are:

+ **glacial trough** with **ribbon lake**: Wastwater, Lake District
+ **hanging valley** (see Figure 4.8): Church Beck flowing into Coniston Water, Lake District
+ **truncated spur** and **glacial trough**: Nant Ffrancon, North Wales
+ **roches moutonnées**: examples in Yosemite National Park, California.

Now test yourself   TESTED ◯

25 Which weathering processes contribute to the formation of an arête?

26 What is a:
   a) bergschrund crevasse
   b) truncated spur?

27 Explain the formation of a roche moutonnée.

28 Explain how ice shapes valleys.

29 What is a ribbon lake?

Answers on p. 262

Revision activity

Devise a revision presentation to explain the formation of erosional landforms and depositional landforms in glaciated landscapes. Limit yourself to ten slides.

Try to move away from PowerPoint if possible, for example Keynote, Google slides.

You could take this further by sharing your presentations in small groups, or in pairs. Each group could present a different aspect, for example one presents erosional features and the other depositional features.

Characteristic glaciated landscapes – erosion features

Erosional landforms produced by glaciers combine to form a classic glaciated landscape, as seen in Snowdonia, North Wales (see Figure 4.9).

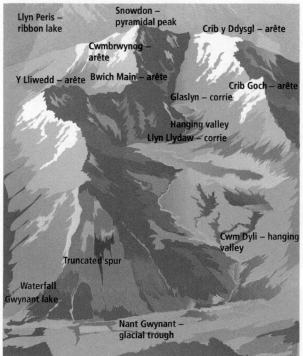

Figure 4.9 Glaciated landscape of Snowdonia

Depositional landforms

Drumlins

Drumlins are smooth, elongated mounds of till (unsorted rock, clay and sand deposited by ice). The long axis runs parallel to the direction of ice movement.

They can be 100 m high and more than 1 km in length.

They are smoothed by abrasion with a steep upstream side and have a gently sloping downside.

In a group, they form a 'swarm'.

Drumlins may be formed by:
+ reshaping previously deposited glacial material
+ accumulation of material around a bedrock obstruction, or
+ thinning of ice in a lowland area leading to deposition of debris.

> **Exam tip**
>
> Drumlins deposited by ice sheets can be extensive, localised and lateral. Medial moraines are on a much smaller scale.

Erratics

Erratics are boulders picked up and carried by the ice (often over many kilometres) to be deposited in areas of a completely different geology.

Moraines

Moraines are landforms created when the debris carried by a glacier is deposited. There are several types:
+ **Lateral:** derived from frost shattering of the valley sides, carried at the edge of a glacier; on melting, a side embankment is formed.
+ **Medial:** the merging of two lateral moraines.
+ **Terminal:** a high mound extending across the valley to mark the maximum advance of the ice sheet.
+ **Recessional:** these mark an interruption in the retreat of the ice.
+ **Push moraines:** these form if the climate deteriorates and the ice advances.

> **Now test yourself**   TESTED ○
>
> 30 Outline two possible ways in which drumlins can form.
>
> 31 Distinguish between recessional and push moraines.
>
> **Answers on p. 262**

Till plains

These are often found behind terminal moraines in low-lying areas. They are wide areas of flat relief where there is a covering of glacial till. Till is usually composed of sand and gravel but this depends on the type of rock over which the ice has travelled. It is usually 30–50 m deep.

Examples of these features include:
+ **drumlin:** Risebrigg Hill, North Yorkshire
+ **erratic:** examples at Ingleborough, Yorkshire Dales
+ **moraine:** Meade Glacier, Alaska
+ **till plain:** south of the Great Lakes, North America.

Check your understanding and progress at **www.hoddereducation.co.uk/myrevisionnotesdownloads**

Characteristic glaciated landscapes – depositional features

Revision activity

Research a photograph of a characteristic glaciated landscape showing depositional features. Label the features that you can identify. List the main processes involved in the formation of these features.

Fluvioglacial landforms

When glaciers melt, meltwater transports large amounts of debris. The resulting meltwater streams often flow at high velocity. A loss of energy due to a decrease in discharge leads to deposition. In this way, fluvioglacial landforms are formed 'downstream' of glaciated areas.

+ **Meltwater channels:** these are classified according to where they flow in relation to the glacier during the glacial period (see Figure 4.10).

1. Supraglacial   2. Marginal   3. Englacial

4. Submarginal   5. Subglacial

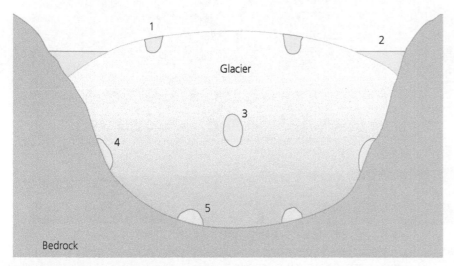

Figure 4.10 Meltwater channel types

+ **Kames:** these are undulating, winding mounds of unevenly deposited sand and gravel. Kame terraces are flat areas formed along the sides of valleys. They follow the direction of ice advance.
+ **Eskers:** these are very long, narrow ridges of sorted, stratified, coarse sand and gravel.
+ **Outwash plains (sandur):** these are deposits by meltwater streams running out from the snout (end) of the glacier. They are composed of coarse material, which is found near to the glacier, and finer clay, which is carried across the plain before being deposited. Within the outwash plain there may be:
   + **braided channels:** meltwater streams that cross the outwash plain which are divided (braided) as the channels become choked with material
   + **kettle holes:** a series of small depressions, formed when blocks of ice washed onto the outwash plain melt and leave a gap in the sediments.

An example of a fluvioglacial landscape with **outwash plains** is Skeidarár Sandur in Iceland.

32 Why do meltwater streams often flow at a high velocity?

33 What are braided channels and why are they common in areas of glacial deposition?

34 How do kettle holes form?

Answers on p. 262

Revision activity

You need to revise in detail how kames, eskers and outwash plains are formed.

Either produce a simplified diagram of a fluvioglacial landscape and annotate it (see Figure 4.11) or create a table and complete it with revision notes (feature, description and processes) – see Table 4.4, eskers has been done for you. You should also refer to your class notes to cover the processes in detail.

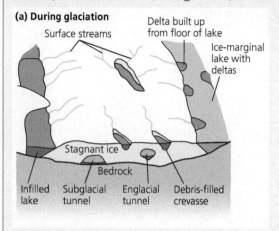

(a) During glaciation

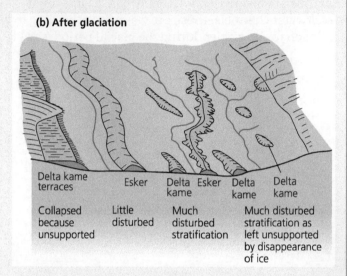

(b) After glaciation

Figure 4.11 Formation of kames and eskers

Table 4.4 Revision of fluvioglacial landforms

Feature	Description	Processes
Esker	Long narrow ridge composed of layered (stratified) sand and gravel	✦ Subglacial streams carry large amounts of debris. ✦ Pressure is released at the snout (end) of the glacier and this causes deposition in the meltwater streams. ✦ As the glacier retreats, the point of deposition will move backwards.
Kames		
Outwash plains		

Characteristic fluvioglacial landscapes

Many areas on the edge of warm-based glaciers develop distinctive fluvioglacial landscapes. Features are summarised in Figure 4.12.

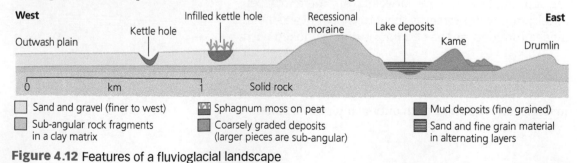

Figure 4.12 Features of a fluvioglacial landscape

Check your understanding and progress at **www.hoddereducation.co.uk/myrevisionnotesdownloads**

Periglacial landforms

Periglacial refers to landscapes that are not actually glaciated but are exposed to very cold conditions, such as the tundra landscapes of Russia, Alaska and Canada.

Patterned ground

Patterned ground is a landform reflecting the repeated cycles of freezing and thawing of the active layer (see Figure 4.13). Rock particles are distributed in a system of polygons and circles. A process of **frost heave** (expansion of the volume of soil as ice crystals form) pushes larger stones to the surface and, due to the camber, stones move sideways.

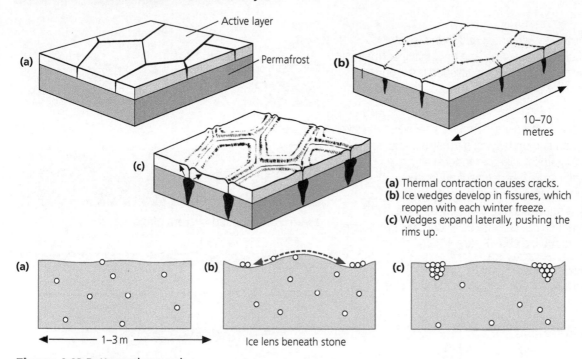

(a) Thermal contraction causes cracks.
(b) Ice wedges develop in fissures, which reopen with each winter freeze.
(c) Wedges expand laterally, pushing the rims up.

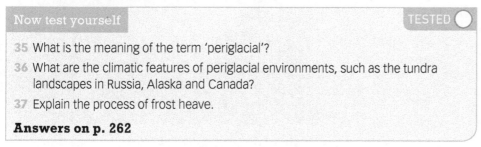

Ice lens beneath stone

Figure 4.13 Patterned ground

Now test yourself

TESTED ◯

35 What is the meaning of the term 'periglacial'?

36 What are the climatic features of periglacial environments, such as the tundra landscapes in Russia, Alaska and Canada?

37 Explain the process of frost heave.

Answers on p. 262

Ice wedges

These are narrow, frost-formed cracks in the upper layers of the ground, which fill with ice. They can be up to 10 m in depth.

Pingos

Pingos are dome-shaped, ice-cored mounds of earth. There are two theories on the formation of pingos – see Table 4.5 and Figure 4.14.

Table 4.5 The formation of pingos

Closed system

Famously found on the Mackenzie Delta area of the Northern Territories, Canada, so known as Mackenzie Delta type.

Stage 1
+ Generally found in areas of continuous permafrost.
+ Develop beneath lakebeds.
+ The growth of the ice core is hydrostatic.
+ Deep lakes (over 2 m) may remain unfrozen in winter.
+ The permafrost layer at the lakebed is insulated from the cold and thaws.

Stage 2
+ An area of unfrozen, waterlogged ground is now sandwiched between the lake and underlying permafrost.
+ The lake may begin to drain; the lakebed is no longer insulated, so the waterlogged bed begins to freeze.

Stage 3
+ Due to localised differences in pressure between the lake, freezing lakebed and underlying permafrost, the newly freezing water gathers together to form an ice lens that expands, pushing the lakebed sediments above it up into the classic dome shape.
+ The ice lens continues to grow as long as there is still unfrozen ground in the lakebed as a source of pressurised water to add to the ice core (Figure 4.14).

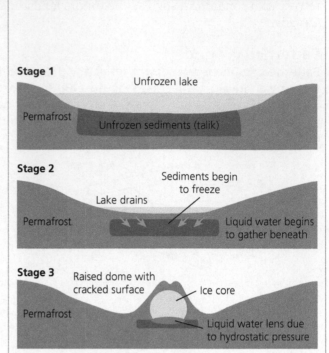

Figure 4.14 Closed-system pingo

Open system

Common in Greenland and Alaska, so known as East Greenland type.
+ Generally found in areas of discontinuous permafrost.
+ Found in valley bottoms.
+ The growth of the ice core is hydraulic.
+ Water is able to seep into the upper layers of the ground and flows from higher surrounding areas under artesian pressure.
+ Water accumulates in flat, low-lying areas between the upper layers of permafrost or soil and frozen ground beneath the water, and then freezes.
+ The freezing ice core expands, thus doming the overlying layers into the classic pingo shape.
+ They grow as pressurised water continues to flow in from their surroundings (Figure 4.15).

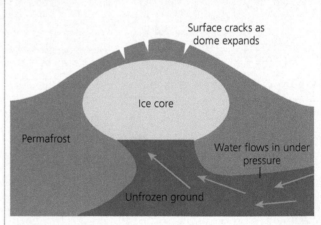

Figure 4.15 Open-system pingo

Blockfields

Freeze–thaw action produces large amounts of scree to form **scree slopes**. In flat areas, expansive areas of angular boulders form blockfields.

Solifluction

This is summer melt of water in the upper layers of permafrost, which leads to large amounts of water that cannot drain away due to the permafrost. The lubrication means that soil is moved on the most gentle of slopes.

Lobes

Where solifluction forms on steeper slopes, tongue-like lobes extend down the slope. They can be 50 m wide and 5 m high.

Terracettes

These are narrow steps with a small tread (tens of centimetres), which run parallel to the contours of a slope.

> **Now test yourself**
>
> 38 State two differences between closed- and open-system pingos.
>
> 39 What is solifluction?
>
> **Answers on p. 262**
>
> TESTED ◯

Check your understanding and progress at **www.hoddereducation.co.uk/myrevisionnotesdownloads**

Thermokarst

When ice melts within permafrost, depressions known as thermokarst form in the ground. They are the result of temperature change, not erosion or weathering.

Examples of these features are:
+ patterned ground: Barrow, Alaska
+ pingo: examples in northern Canada
+ blockfield: Scafell Ranges, Lake District
+ solifluction sheet: Ogilvie Mountains, Canada.

Characteristic periglacial landscapes

Figure 4.16 shows how the landforms outlined above can combine to form a characteristic periglacial landscape.

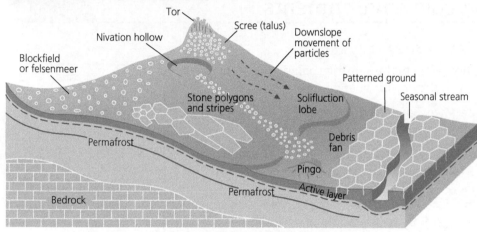

Figure 4.16 Features of periglacial landscapes

Revision activity

Make a large copy of Figure 4.16 and for each landform:
+ add one brief sentence giving a description of the landform
+ list the process(es) contributing to its development.

Example:

Blockfield

Description: angular boulders lying in flat, expansive areas.

Process: freeze–thaw weathering.

The relationship between process, time, landforms and landscapes in glaciated settings

Revision activity

There is a range of processes relevant to the cold environments covered in this unit. Make a summary list of those which are effective in glacial, periglacial and fluvioglacial environments.

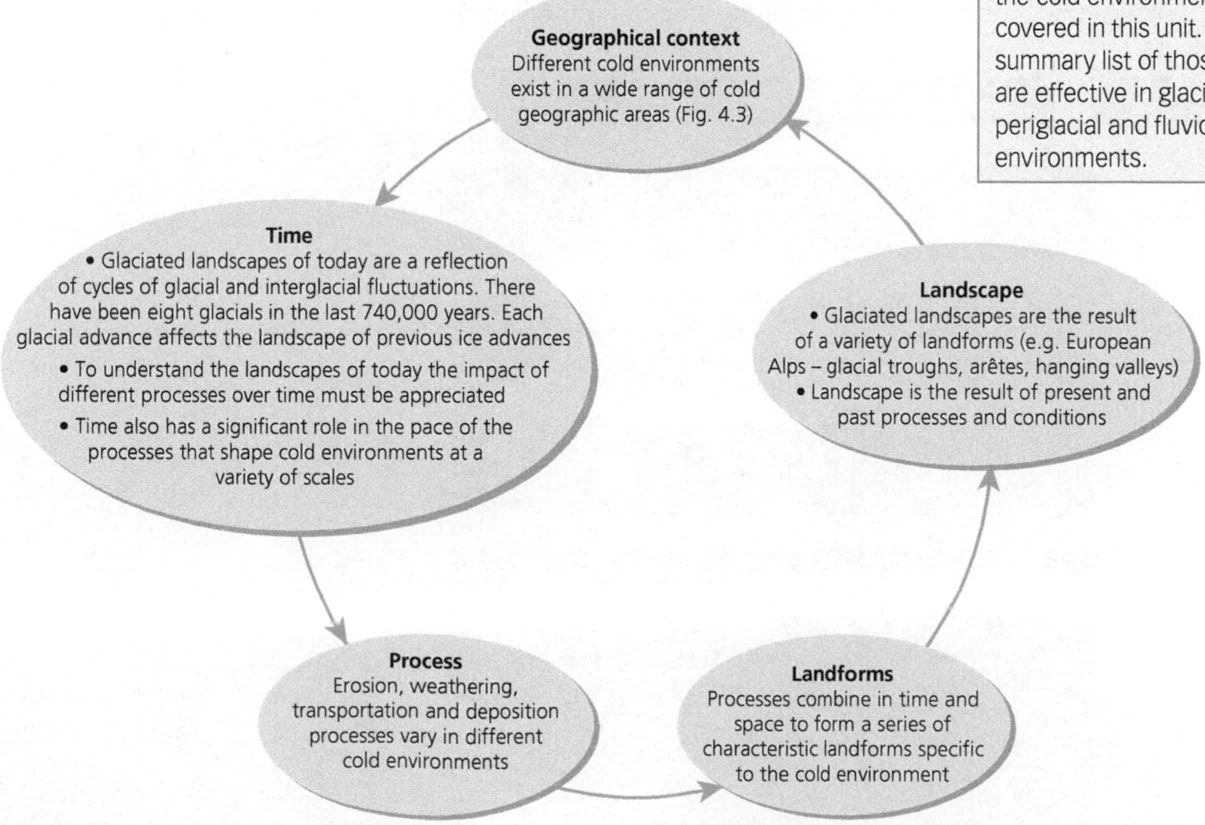

Geographical context
Different cold environments exist in a wide range of cold geographic areas (Fig. 4.3)

Time
• Glaciated landscapes of today are a reflection of cycles of glacial and interglacial fluctuations. There have been eight glacials in the last 740,000 years. Each glacial advance affects the landscape of previous ice advances

• To understand the landscapes of today the impact of different processes over time must be appreciated

• Time also has a significant role in the pace of the processes that shape cold environments at a variety of scales

Landscape
• Glaciated landscapes are the result of a variety of landforms (e.g. European Alps – glacial troughs, arêtes, hanging valleys)
• Landscape is the result of present and past processes and conditions

Process
Erosion, weathering, transportation and deposition processes vary in different cold environments

Landforms
Processes combine in time and space to form a series of characteristic landforms specific to the cold environment

Figure 4.17 Glaciated settings; the links between process, time, landforms and landscapes

My Revision Notes: AQA A-level Geography Second Edition

Revision activity

Using the table headings below, summarise a checklist of the characteristic landforms in each of the four landscapes: glaciated erosional, glaciated depositional, fluvioglacial and periglacial. A couple of examples have been given for you.

Erosional landforms in glaciated landscapes	Depositional landforms in glaciated landscapes	Landforms of fluvioglacial landscapes	Landforms of periglacial landscapes
Examples: + Corries	Examples: + Erratics	Examples: + Kames	Examples: + Patterned ground

Exam tip

It is key with the physical geography units to understand how landforms combine to form characteristic landscapes.

Human impacts on cold environments

The concept of environmental fragility

For natural environments, **fragility** refers to the sensitive balance between non-living (climate, geology, soils) and living (flora and fauna) components.

Human activity can upset this balance. Some environments are robust while others are more vulnerable to climate change and human activity.

Cold environments are classified as fragile environments due to the following:
+ Slow rates of soil formation and plant growth mean that ecosystems are slow to recover after change.
+ Lack of biodiversity and simple food webs mean that the decline of a single species can destabilise the ecosystem.
+ Thawing of the permafrost can lead to geomorphological, hydrological and ecological change.

Flora The plants of a particular region, habitat or geological period.

Fauna The animals of a particular region, habitat or geological period.

Human impacts on fragile cold environments over time

Figure 4.18 shows how human activity has had an impact on fragile cold environments.

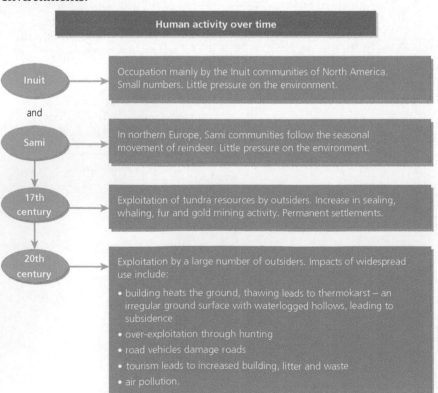

Figure 4.18 Timeline of human activity in fragile cold environments

Check your understanding and progress at **www.hoddereducation.co.uk/myrevisionnotesdownloads**

Human impacts on fragile, cold environments at a variety of scales

Cold environments offer a range of opportunities for human populations (see Table 4.6). They include traditional occupations of hunting, fishing, caribou and reindeer herding, and fur trade.

Innovations have overcome some of the problems of permafrost to allow building construction and improved infrastructure.

Table 4.6 Economic activity in cold environments

Cold environment	Description of economic activity
Alaska	Oil, mining, timber, fishing and tourism
European Alps	Winter sports tourism – high snowfall, steep slopes and high standards of services and facilities
	Summer tourism – warm summers, potential for hiking, climbing and cycling
	Hydroelectric power (HEP) potential in glacial troughs with hanging valleys and differential heights
Siberia	Mining of iron, silver, copper and gold. Also, resources of coal, oil and gas
	Vast forests provide for a large timber industry

Challenges to human activity include:
+ harsh climate (very low temperatures, short summers, low precipitation, thin, poorly developed soils, frozen ground, blizzards, persistent snow cover)
+ shortage of skills and labour
+ remoteness
+ lack of permanent jobs
+ limited education opportunities past secondary level.

Recent and prospective impact of climate change

The majority of climate scientists agree that the Earth's climate is changing. Recent decades have seen a pattern of accelerated warming of global temperatures. The Earth's climate changes naturally but the concern is that human activity is furthering the process of global warming.

Figure 4.19 shows the recent impacts of climate change in cold environments.

Figure 4.19 Recent impacts of climate change in cold environments

Now test yourself

40 Outline how time affects glacial landscapes.
41 Why are cold environments described as fragile?

Answers on p. 262

TESTED ◯

Making links

See Chapter 1, pages 30–31 for more detail on the causes and impacts of climate change.

Exam tip

Remember to draw on knowledge of climate change from other parts of the course, for example an explanation of the greenhouse effect.

Revision activity

Using Figures 4.19 and 4.20, make bullet point revision notes on three impacts of climate change on cold environments that have been observed and also on three predicted future changes.

Figure 4.20 shows the predicted future changes in cold environments.

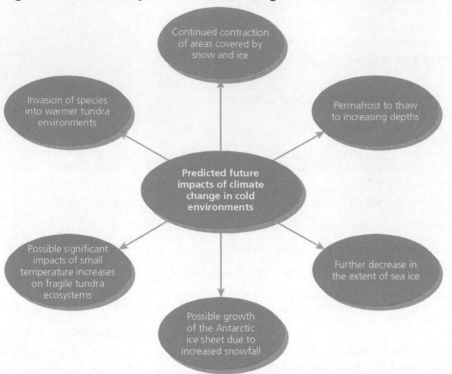

Figure 4.20 Predicted future impacts of climate change in cold environments

Present management of cold environments

Present management of cold environments includes the protection of Antarctica as one of the global commons.

Table 4.7 outlines how Antarctica is managed.

Table 4.7 Management of Antarctica as a global commons

Governance/protection	How it protects Antarctica
The Antarctic Treaty System (1959)	+ Military and nuclear activity is banned. + Scientific research is protected. + Rules are established to manage tourism and research activities.
The Madrid Protocol (1998)	+ All activity related to mineral resources other than scientific research is banned. + All activities require an environmental assessment. + Arbitrates international disputes about Antarctica.

> **Making links**
>
> For more information on Antarctica as one of the global commons, see Chapter 7 page 159.

> **Exam tip**
>
> There are also pressures regarding the future sustainability of cold environments in contrasting geographical locations. Be prepared to adapt your understanding to unfamiliar examples.

Alternative possible futures for management

There is currently debate about whether to renew the Antarctic Treaties in light of pressures for future development.

Arguments regarding possible future development in Antarctica include:
+ Global energy and mineral sources are becoming depleted. Antarctica is believed to have reserves of coal, oil and precious metals.
+ Tourism numbers to the Antarctic are high: 2700 visitors in 2008–09. Is there scope to increase tourism as an economic activity?
+ Fishing stocks in some parts of the world are severely depleted, while the Southern Ocean has plentiful stocks.
+ Bioprospecting is a growing area of scientific research. Many companies are keen to investigate the biochemical resources of Antarctica's flora and fauna.

Check your understanding and progress at **www.hoddereducation.co.uk/myrevisionnotesdownloads**

Now test yourself

TESTED ○

42 State three ways in which Antarctica is presently managed.

43 Give two examples of pressure to develop the resources of Antarctica.

Answers on p. 263

Revision activity

Make revision notes on your local case study, focusing on the aims and outcomes of fieldwork and on the features of the glaciated landscape observed. Illustrate glacial processes, summarise landscape outcomes and engage with field data.

Case studies

A glaciated environment at a local scale, e.g. the Helvellyn area of the Lake District

This case study should:

+ illustrate and analyse fundamental glacial processes and their landscape outcomes
+ engage with field data.

What do I need to know?	Content and suggested revision methods
Glacial processes and their landscape outcomes	**Processes** The processes at work on the chosen landscape are time as well as place specific, so you can set out your revision notes as a timeline, including full explanations of both **glacial** and **interglacial processes** (weathering and mass movement). **Glacial maximum** • Pleistocene glaciation has modified the landscape • Glacial advance and retreat 11× during the last 800,000 years • Last glacial maximum in NW England about 22,000 years BP **Glacial advance** • Abrasion • Plucking **Glacial retreat** • Glacial deposition processes • Meltwater erosional processes **Landscape outcomes** You can revise landscape features by producing a series of labelled diagrams, by combining landscape features in a sketch or by tabulating information. <table><tr><th>Landscape feature</th><th>Description</th><th>Named location</th></tr><tr><td>Corries</td><td>150 + in the Lake District Most on the east side of the Helvellyn Range 600 m+ Many corries face NE Snow accumulates in cold periods Arêtes also a feature of the landscape</td><td>Brown Cove. Arêtes – Swirral Edge and Striding Edge</td></tr></table> Other landscape features to revise include: + glacial troughs + moraines + ribbon lakes + roche moutonnées.

A contrasting glaciated landscape from beyond the UK, or example the Sapmi region, northern Europe

This case study should:

+ illustrate and analyse how it presents challenges and opportunities for human occupation and development
+ evaluate human responses of resilience, mitigation and adaptation.

What do I need to know?	Content and suggested revision methods
Challenges and opportunities for human occupation and development	It is a good idea to organise these revision notes in a table as both challenges and opportunities can be further subdivided into economic, social and environmental. For example:

Economic	Social	Environmental
CHALLENGES		
Commercial logging threatens herding land	Cultural traditions threatened	Lakes and rivers make long-distance travel difficult
OPPORTUNITIES		
Economic needs are low as reindeer provide food, clothing and utensils	Strong links in family groups and longstanding traditions	Diversity of environment allows breeding of different reindeer

What do I need to know?	Content and suggested revision methods
Human responses of resilience, mitigation and adaptation	You can organise your notes as bulleted points or on a star diagram with colour coding for resilience, mitigation and adaptation. For example: Adaptation + Change from a nomadic way of life to farm reindeer in a more commercial, settled way. Resilience + Young people have stayed and so keep the population viable. Mitigation + International organisations and national governments work to protect indigenous people and mitigate against negative environmental change.

Exam practice

1 Outline the main features of periglacial landscapes. [4]
2 Explain how cold environments are being managed. [4]
3 Compare and contrast the characteristic features of open- and closed-system pingos. [6]
4 Assess the challenges to human activity in cold environments. [6]
5 Assess present-day human impact on cold environments. [6]

Answers and quick quizzes online

Exam skills

Opportunities to practise geographical skills within this topic include:
+ observation skills in fieldwork
+ collection, manipulation, presentation and analysis of primary and secondary fieldwork data (including quantitative and qualitative sources)
+ analysis of specific graphs, such as climate graphs
+ interpretation of geospatial data (e.g. global maps showing glaciated areas, soil types, climate data, tundra environments)
+ data manipulation and interpretation of, for example, glacial budgets, rates of glacial erosion or glacier retreat.

Summary

+ The systems approach can be applied to the study of glaciers, with glacial budget and mass balance being key concepts.
+ There is a range of 'cold environments' with a widespread global distribution and varying characteristics of climate, soils and vegetation.
+ Glacier movement takes place through flows and slides. The movement may be rotational, extensional and compressional.
+ Weathering processes are important to loosen rock in situ before further processes take place.
+ Classic landscapes are formed, which include a range of erosional and depositional landforms.
+ Deposition can occur directly from the ice (for example moraines) or from meltwater (fluvioglacial).
+ Periglacial environments are not glaciated – landforms are the result of permafrost, ground ice and mass movement processes.
+ Indigenous groups have lived sustainably in cold environments for thousands of years. Today, there is an increasing level of human impact and pressure for further economic development.
+ There is growing debate as to whether current protection treaties should be adapted to allow development or removed altogether.

Check your understanding and progress at **www.hoddereducation.co.uk/myrevisionnotesdownloads**

5 Hazards

The concept of hazards in a geographical context

REVISED

Nature, forms and potential impacts of hazards

A **hazard** is a threat (natural or human) that has the potential to cause loss of life, injury, property damage, socio-economic disruption or environmental degradation. Natural hazards exist at the interface between physical and human geography.

There are three broad groups of natural hazards:

+ **Geophysical:** caused by movements of the Earth. They occur with minimum warning and include earthquakes, volcanic eruptions and tsunamis, landslides and avalanches. They are difficult to predict and impossible to stop.
+ **Atmospheric:** weather-related disasters, for example hurricanes, tornadoes, extreme heat and extreme cold weather. There will usually be some advance warning but the unpredictable nature of weather means that nothing can be done to stop the disasters that result.
+ **Hydrological:** water-related hazards, for example floods, mudslides and landslides. These are usually the consequence of extreme weather events, or are a secondary effect of other natural disasters.

> **Natural hazard** Events that are perceived to threaten people and the built and natural environments.

Making links

There are links between climate change and an increase in extreme atmospheric and hydrological events – see Chapter 1 page 27.

There are also links between climate change and an increase in extreme weather events such as tropical storms and wildfires – see Chapter 1, page 25.

Exam tip

Magnitude is only one of several factors that will affect the impact of a hazard. Overall, the relationship may be quite weak. Other factors could include the level of prevention and protection measures put in place, the level of preparedness, population density and building quality.

Hazards can also be characterised by their:

+ **magnitude** (for instance, measuring an earthquake on the Richter scale)
+ **frequency** or **how often a disaster event of a certain size occurs** (for example, a flood of 1m may be an annual event but one of 2m may happen only every 10 years)
+ **duration** (periods of extreme heat, for example, can last for weeks; an earthquake may be over in minutes)
+ **spatial concentrations** (such as the Ring of Fire, a concentration of earthquakes and volcanoes in the Pacific Basin)
+ **speed of onset** (perhaps the lag time on a flood hydrograph)
+ **temporal spacing** (for example, the seasonal regularity of cyclones).

> **Making links**
>
> In Chapter 1, page 17, there is further information on flooding and flood hydrographs.

The potential impacts of hazards vary greatly over time and space and are dependent on a number of environmental, economic and social factors. Mitigation, experience, perception, physical setting, technology, wealth and place vulnerability will all determine the potential impact of a hazard.

Hazard perception

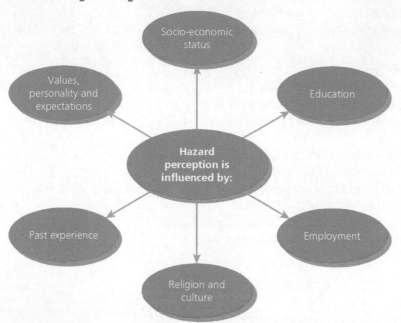

Hazard perception The way in which people view the threat of the hazard.

Figure 5.1 Hazard perception

People can perceive natural hazards in the following ways – each is dependent on economic and socio-cultural determinants:

+ Fatalism – there is an acceptance of the hazard, losses are accepted as inevitable and people remain where they are.
+ Adaptation – there is a positive view of prediction, prevention and protection, which will depend on the economic status of the area.
+ **Fear** – people feel vulnerable, they cannot live with the threat and they move away.

Fatalism The hazard is inevitable and people cannot influence the outcome.

Adaptation Attempts/ changes made to live with a hazard.

Now test yourself   TESTED ◯

1 Define a hydrological hazard and give two examples.
2 What is the meaning of the term 'mitigation'?
3 What factors determine the impact of a hazard?
4 What factors influence hazard perception?

Answers on p. 263

Characteristic human responses

Responses occur at different levels: individual, community, national government and international.

+ **Resilience** is the sustained effort of communities to respond to and withstand the effects of hazards.
  + A process known as integrated risk management incorporates the identification of a hazard, risk analysis and a risk reduction plan with monitoring and review.

Check your understanding and progress at **www.hoddereducation.co.uk/myrevisionnotesdownloads**

- Hazards are also managed by the integration of prediction, prevention and protection plans. Prediction is not always possible scientifically but by careful monitoring, warnings can be issued.
+ **Natural hazards** cannot be prevented, but some of the dangerous secondary impacts can be controlled – for example, lava flows can be diverted. Protection aims to minimise the impact of a hazard event through, for example, adaptations to the built environment, such as erecting earthquake-proof buildings and sea defences.
+ **Risk sharing** involves public education and awareness of the measures available to reduce impact, such as emergency responses and evacuation procedures.

All of the above can be put together in a **hazard management cycle**, as seen in Figure 5.2.

Making links

The general concepts outlined under the heading 'Characteristic human responses' must be applied to each hazard type within this unit:
+ plate tectonics (page 101)
+ storm hazards (page 112)
+ wildfires (page 115).

Figure 5.2 The hazard management cycle

The Park model of human response to hazards

The Park model shows that hazards have varying impacts over time:
+ before the disaster
+ when the event happens and post-event relief (rescue)
+ rehabilitation (restoring the functioning of public services) and reconstruction (rebuilding the public and economic system, replacing infrastructure and governance), as shown in Figure 5.3.

Exam tip

Subdividing responses to hazards by scale, for example individual to community, national and international, gives an excellent structure to extended answers in examinations.

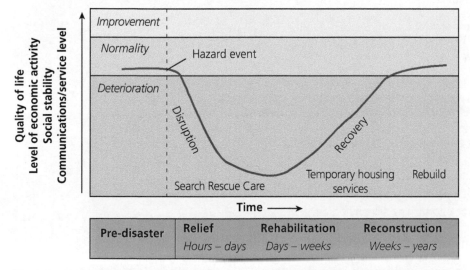

Figure 5.3 The Park model of human responses to hazards

My Revision Notes: AQA A-level Geography Second Edition

Plate tectonics

REVISED ◯

Earth structure and internal energy sources

Earth's structure

The Earth's cross section is shown in Figure 5.4. It comprises a sequence of shells – see Table 5.1.

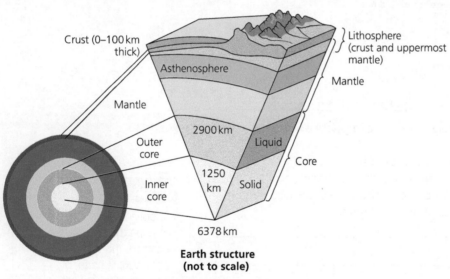

Crust (0–100 km thick)
Asthenosphere
Mantle
2900 km
Outer core
1250 km
Inner core
6378 km

Lithosphere (crust and uppermost mantle)
Mantle
Liquid
Core
Solid

Earth structure (not to scale)

Figure 5.4 The structure of the Earth

Table 5.1 The layers of the Earth

Layer	Description
Core	+ The centre of the Earth + An iron–nickel mass that gives the Earth its magnetic field + The inner core is 1250 km thick + The outer core is liquid and is 2200–2900 km thick
Mantle	+ Accounting for more than 80% of the volume of the Earth + It consists of semi-solid rock containing silicon and oxygen + It is 2900 km deep + The upper part of the mantle – the **asthenosphere** – has plastic properties that allow it to flow under pressure

Crust	+ The outer shell consisting of:
	  + oceanic crust (solid) composed of dense basalt rock – average 5 km thick
	  + continental crust (solid), mainly granite, which is less dense than basalt, averaging 30 km thick. Continental crust can be up to 100 km deep under major mountain ranges

Asthenosphere The layer of upper mantle extending 100 km to 300 km, slow-flowing and viscous.

The crust, together with the immediate underlying mantle, makes up the **lithosphere**.

Check your understanding and progress at **www.hoddereducation.co.uk/myrevisionnotesdownloads**

Internal energy sources

The Earth's internal heat source provides the energy for plate tectonic motion and, consequently, earthquakes and volcanic eruptions. This internal heat energy has accumulated rapidly over the years due to two main processes:

+ conversion of gravitational energy
+ radioactive decay of unstable isotopes.

Plate tectonic theory

The theory of plate tectonics suggests the following:

+ The lithosphere is divided into seven large and three smaller tectonic plates.
+ New lithosphere or crust is added to constructive plate margins.
+ Old crust is moved laterally away from these constructive margins, a process known as sea floor spreading, and is eventually destroyed in subduction zones or destructive plate margins.
+ This constant recycling, resulting in a slow movement of the continents across the Earth's surface, is known as continental drift.
+ Computer models show that the cooling rock at constructive plate margins exerts a force on spreading lithospheric plates that could help drive their movement. This force is called ridge push, or sometimes gravitational sliding.
+ At destructive plate margins, one plate is denser and heavier than the other plate. The denser plate subducts beneath the plate that is less dense. Slab pull is the force that the sinking plate exerts on the rest of the plate.
+ There is disagreement among scientists as to which processes are most effective in plate movement:
    + Many consider slab pull to be a stronger force than either ridge push or mantle convection.
    + Others believe ridge push to be the most significant force.
    + Most scientists, however, do believe that convection currents play only a supporting role in the movement of plates.

> **Tectonic plate** Rigid sections of the Earth's crust that float on the upper mantle and move relative to one another.
>
> **Sea floor spreading** Theory that the ocean floor is moving away from the mid-Atlantic ridge and crust must be destroyed elsewhere.

Plate movement

Scientists now believe that convection currents play only a small part in dynamic models of plate movement (ridge push and slab pull). Plates move as part of a gravity-driven system (see Figure 5.5).

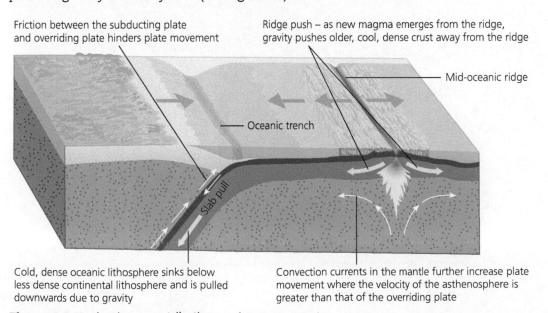

Friction between the subducting plate and overriding plate hinders plate movement

Ridge push – as new magma emerges from the ridge, gravity pushes older, cool, dense crust away from the ridge

Mid-oceanic ridge

Oceanic trench

Slab pull

Cold, dense oceanic lithosphere sinks below less dense continental lithosphere and is pulled downwards due to gravity

Convection currents in the mantle further increase plate movement where the velocity of the asthenosphere is greater than that of the overriding plate

Figure 5.5 Mechanisms contributing to plate movement

My Revision Notes: AQA A-level Geography Second Edition

Plate margins

The lithospheric plates interact with one another at their margins. It is along these margins that most volcanic and seismic activity occurs. There are three types of plate margin:

+ destructive
+ constructive
+ conservative.

Revision activity

Using the key words below, create a flow diagram that summarises plate tectonic theory. You can add detail from your own notes.

+ Lithosphere
+ Gravity-driven system of plate movement
+ Destructive plate margins
+ Constructive plate margins
+ Conservative plate margins
+ Ridge push
+ Slab pull

Destructive, constructive and conservative plate margins

Destructive plate margins

Destructive plate margins are also referred to as **subduction zones** and it is here that lithosphere is destroyed.

Subduction occurs when two tectonic plates converge. The older, denser plate is subducted below the less dense plate, as in Figure 5.6. Destructive plate margins are the location of volcanoes and earthquakes.

Revision activity

Practise simple diagrams of the different types of plate boundary and fully annotate them with an explanation of processes and landforms.

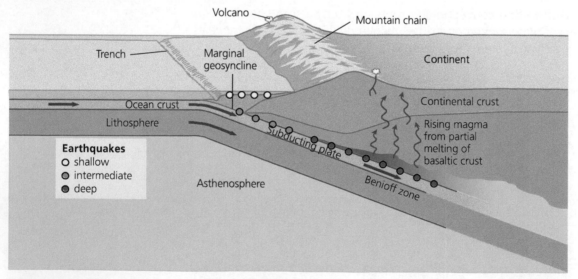

Figure 5.6 A destructive plate margin

Where two oceanic plates converge, subduction forms an island arc (see Figure 5.7).

Island arcs

+ Island arcs form during subduction. The descending plate encounters hotter surroundings, which together with frictional heat, cause the plate to melt.
+ As the subducted material is less dense than the surrounding asthenosphere, it rises to the surface as plutons of magma.

Island arc A chain of volcanic islands which form during subduction.

Magma Molten rock found beneath the Earth's surface.

Check your understanding and progress at **www.hoddereducation.co.uk/myrevisionnotesdownloads**

+ On reaching the surface, these form explosive volcanoes.
+ If the eruptions take place offshore, a line of volcanic islands forms – such as the island arc of the Mariana Islands.

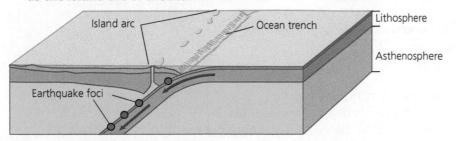

Figure 5.7 Formation of an island arc

Fold mountains
+ Where an oceanic plate and continental plate converge, fold mountains form.
+ Where two continental plates meet at destructive plate margins, there is not much subduction due to the low density of continental crust. Instead, their edges are forced up into fold mountains, for example the Himalayas at the Indo-Australian Plate and Eurasian Plate margin.

Ocean trenches
+ These form at destructive plate margins between oceanic plates, or between oceanic and continental plates. In both cases, one plate subducts beneath the other.
+ An example of the former is the Mariana Trench – formed where the Pacific Plate subducts beneath the smaller Philippine Plate.
+ An example of the latter is the Peru–Chile Trench – formed where the heavy, dense Nazca Plate subducts under the South American Plate.

Constructive plate margins
New crust forms at constructive plate margins where rising plumes of magma from the upper mantle stretch the crust and lithosphere (see Figure 5.8). The resulting intense tectonic activity builds:
+ **mid-ocean ridges** – submarine mountain ranges
+ **rift valleys** – parallel faults (Figure 5.8).

Figure 5.8 A constructive plate boundary

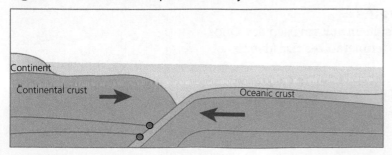

● Earthquake foci

Figure 5.9 Outline diagram for the revision of the processes and landforms at an oceanic–continental destructive plate boundary

Now test yourself

10 How do island arcs form?

11 How have the Himalayas formed?

Answers on p. 263

TESTED ○

5 Hazards

Plumes of magma Hot columns of magma rising from deep within the Earth.

Revision activity

Draw a series of simple diagrams with annotations to summarise the processes and landforms associated with each type of plate boundary. Figure 5.9 gives an outline example for an oceanic and continental plate (annotations and labels of landforms need to be added).

My Revision Notes: AQA A-level Geography Second Edition

Ocean ridges
+ These are formed when plates move apart in oceanic areas.
+ The space between the plates is filled with basaltic lava from below to form a ridge.
+ Volcanoes also exist along such ridges and may rise above sea level, for example Surtsey, south of Iceland.

Rift valleys
+ These are formed when plates move apart in continental areas.
+ Sometimes the brittle crust fractures as sections of it move and areas of crust drop down between parallel faults to form the valley, for example the East African Rift Valley.

Conservative plate margins
At a conservative plate margin, two plates slide past each other.

The movement can be violent and an additional build-up of pressure, which eventually gives way, results in powerful earthquakes. There is no volcanic activity.

An example is the San Andreas fault between the Pacific and North American plates.

Magma plumes
Magma plumes are columns of magma rising through the mantle. On reaching the base of the lithosphere, the magma spreads and the overlying lithosphere is pushed up and stretched. Temperatures rise to above melting point and these magma chambers feed volcanoes; the area is known as a **hot spot**.

> **Now test yourself** TESTED ○
>
> 12 Distinguish between ocean ridges and rift valleys.
> 13 What is a hot spot?
> 14 List the differences between oceanic and continental crust.
> 15 How is a rift valley formed?
> 16 What are the differences between the volcanic activity at constructive and destructive plate margins?
>
> **Answers on p. 263**

Volcanic hazards

REVISED ○

The nature of vulcanicity and its relation to plate tectonics
The majority of volcanic activity is associated with plate margins.

Vulcanicity at destructive plate margins
+ Subduction at destructive plate margins results in volcanic activity. Once oceanic crust dips below continental crust, temperatures rise (due to increased depth) and melting occurs.
+ The resulting magma moves slowly to the surface where it erupts through volcanoes and fissures as viscous, thick **andesitic lava** and **tephra** (ash).
+ The viscous magma traps steam and other gases, creating violent explosions and eruptions.

Vulcanicity at constructive plate margins
+ At constructive plate margins, tension in the crust and lithosphere reduces pressure and allows magma to flow to the surface.

- Lava, tephra and hot gases erupt through volcanoes and fissures.
- Most eruptions are on the ocean floor as constructive plate margins are found at mid-ocean ridges.
- Differences from the volcanic activity at destructive margins include:
    - the lava is **basalt** (low viscosity, flows long distances before cooling and solidifying) rather than andesitic
    - eruptions are less violent – **effusive** – because gases escape easily from the basalt.

> **Exam tip**
>
> The key to understanding eruption behaviour at destructive and constructive plate boundaries is the viscosity of the magma and the extent of the release of gases and steam.

Forms of volcanic hazard

Volcanoes have a range of primary and secondary effects.

Table 5.2 Primary and secondary effects of volcanic hazards

Primary effects	
Tephra	Volcanic bombs and **ash** are ejected into the atmosphere
Pyroclastic flows – nuées and ardente	Gas and tephra, which are extremely hot (over 800°C), flow down the sides of the volcano at speeds of 700 km per hour
Lava flows	Flows or streams of molten rock pour from an erupting vent. The speed at which lava moves depends on: + the type of lava + its viscosity + the steepness of the ground + whether the lava flows as a broad sheet, through a confined channel or down a lava tube
Volcanic gases	Carbon dioxide, carbon monoxide, sulphur dioxide and chlorine escape through fumaroles (openings in or near a volcano through which hot, sulphurous gases escape)
Secondary effects	
Volcanic mud flows (lahars)	These are a combination of melted snow and ice, rock, sand and volcanic ash. They are capable of flowing at high speeds and over long distances, following valley courses
Flooding	Serious flooding can result when eruptions melt glaciers and ice caps. This frequently occurs in Iceland, where the floods are known as jökulhlaups
Acid rain	Volcanoes emit gases, which include sulphur. This combines with atmospheric moisture to form acid rain

> **Now test yourself**　　　　　　　　　　　　　TESTED ◯
>
> 17 What are pyroclastic flows?
> 18 Outline two secondary effects of volcanic eruptions.
> 19 What factors affect the speed of lava flow?
>
> **Answers on p. 263**

Distribution, magnitude and frequency of hazard events

Distribution

The global distribution of active volcanoes is shown in Figure 5.10. The pattern is associated with plate margins and subduction zones around the Pacific 'Ring of Fire'.

Lava Hot molten rock from a volcano.

Primary effects The effects resulting directly from the event, for example lava and pyroclastic flows.

Secondary effects The effects resulting from the impact of the hazard, for example flooding and tsunamis.

Ash Dust-sized particles of rock produced by explosive volcanic eruptions.

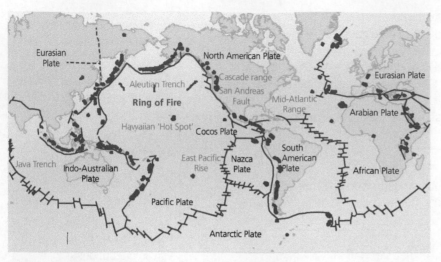

Figure 5.10 Global distribution of active volcanoes

Magnitude

Volcanic eruptions can vary from sluggish to violent explosions. They are divided into two groups:

+ explosive (more violent)
+ effusive (relatively gentle).

Characteristics of each type of explosion vary according to a range of factors, including location, lava type, materials produced and frequency.

The main measurement of magnitude is the **volcanic explosivity index (VEI)**, a logarithmic scale running from 0 to 8; quiet lava produces eruptions rated 0–1, while colossal, violent eruptions score 7–8.

Frequency

Volcanologists determine the frequency of eruptions by studying historical activity. As part of their studies, they collect deposits associated with the volcano.

> **Making links**
>
> See Chapter 7 to understand more on the possible impacts of tectonic hazards on international trade.

> **Frequency** Distribution over time.
>
> **Magnitude** The size of the impact of a hazard event.

> **Revision activity**
>
> From your own notes and possible further research, create a table summarising the characteristics of explosive and effusive eruptions.

Impacts of volcanic activity

Table 5.3 summarises some of the impacts of volcanic hazards.

Table 5.3 Impacts of volcanic hazards

Primary hazards	Secondary hazards	Impacts
+ Pyroclastic flows + Tephra + Lava flows + Ash fallout + Volcanic gases	+ Mudflows + Landslides + Acid rainfall + Flooding	+ Destruction of natural environments and natural ecosystems + Destruction of human environments + Loss of life + Disruption to travel + Damage and destruction to buildings, settlements, property and infrastructure + Disruption to livelihoods + Disruption of communications

All of these impacts will result in economic costs of rescue, rebuilding and repair.

> **Making links**
>
> Volcanic events will have an impact on the character of places – see Chapter 8 page 169 for more.

Check your understanding and progress at **www.hoddereducation.co.uk/myrevisionnotesdownloads**

Risk management

Prediction

It remains difficult to predict volcanic activity. The study of recent history is important, along with an understanding of the type of activity evident. Methods of prediction include monitoring:

+ land swelling
+ changes to groundwater levels
+ chemical composition of groundwater
+ gas emissions
+ expanding cracks
+ looking for shock waves that result from magma moving towards the surface.

Protection

Protection usually refers to preparing for the event.
+ Monitoring can lead to evacuation responses.
+ Risk assessments can lead to a series of alert levels.
+ Land-use planning may follow an assessment of the areas most at risk.
+ It may be possible to divert lava flows away from the built environment with explosives or by using water to solidify lava as it escapes.

A recent volcanic event – impacts and human response

Revision activity

In class you will have studied an example of a recent volcanic event and you should create revision notes about it. The focus should be on the impacts of the event and the human responses. You will also need a brief description of the event – location, date, sequence of events. A table or spider diagram can be used to summarise the impacts and human response. Remember to categorise the impacts as environmental, social, economic and political and also short or long term.

Exam tip

When discussing a hazard event, the impact is the result of exposure and vulnerability of people and installations.

Now test yourself   TESTED

20 State four ways in which volcanic hazards can be mitigated.

Answer on p. 264

Prediction The ability to forecast hazardous events so warnings can be given and action can be taken.

Revision activity

Adapt Table 5.3 to indicate which impacts are social, economic, political and environmental, and which are short-term and long-term impacts.

Making links

Despite the dangers, people continue to take the risk and live in areas of tectonic activity. See Chapter 8 for more on developing a sense of place.

Seismic hazards   REVISED

The nature of seismicity

Seismic waves are vibrations in the Earth's crust. These cause:
+ **earthquakes** – ground shaking (the primary hazard) by the fracturing of rocks
+ sudden movements along **fault lines** (where two tectonic plates meet).

Ground rupture, the visible breaking of the Earth's surface, can also occur.

The precise location of an earthquake within the crust is known as the **focus**.

The point on the surface immediately above the focus is the **epicentre**, where there is the most destruction (see Figure 5.11).

Secondary hazards

+ Shock waves and seismic waves:
    + The shifting rock in an earthquake causes shock waves – called seismic waves – to spread through the rock in all directions.
    + There are two main types of seismic wave associated with earthquakes:
        – P-waves, which travel at around 6–7 km h^{-1}, parallel and through solids and liquids
        – S-waves, which travel sideways at 2.5–4 km h^{-1} and cannot travel through liquid.
+ Tsunamis:
    + A tsunami is a giant sea wave generated by shallow-focus underwater earthquakes, volcanic eruptions and large landslides into the sea.
    + Tsunamis have a long wavelength (often over 100 km) and low wave height (under 1 m) in the open ocean.
    + They travel quickly (speeds of more than 700 km h^{-1}) but, on reaching the shallow water bordering land, they increase in height.
    + A wave trough forms in front of the tsunami where sea level is reduced – this is called a drawdown. Behind this comes the tsunami itself, sometimes as high as 25 m or more.
+ **Liquefaction:** this is when violently shaken soils with a high water content lose their mechanical strength and become fluid.
+ **Landslides:** this is when there is slope failure as a result of the ground shaking.

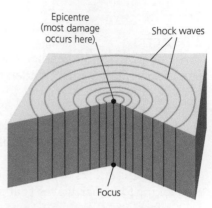

Figure 5.11 The focus and epicentre of an earthquake

Now test yourself

21 Explain the term 'liquefaction'.

Answer on p. 264

TESTED

Spatial distribution

Plate boundaries

The majority of earthquakes occur along plate boundaries, the most powerful at destructive margins and subduction zones (see Figure 5.12).

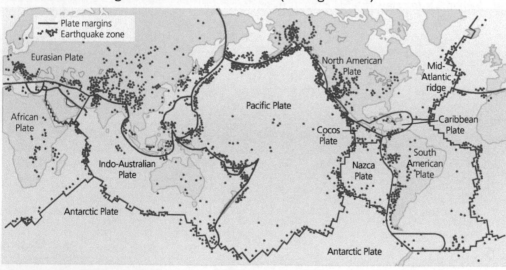

Figure 5.12 Global distribution of earthquakes

+ There is a clear line of earthquakes along the centre of the Atlantic Ocean between the African and American plates and around the Pacific Ocean at the edge of the Pacific Plate.
+ Broad linear chains occur, for example on the west coast of South America.
+ Conservative plate boundaries, where plates are sliding past each other (for example California's San Andreas Fault line), give a relatively narrow band of earthquakes.

Other places where earthquakes occur

+ **Old fault lines:** some earthquakes occur away from plate boundaries and are associated with the reactivation of old fault lines.

Check your understanding and progress at **www.hoddereducation.co.uk/myrevisionnotesdownloads**

+ **Human activity:** it has been suggested that human activity can lead to minor earthquakes, for example the building of large reservoirs which put pressure on the underlying rocks, or deep mine subsidence, or maybe fracking (hydraulic fracturing of rock to release gas).
+ **Hot spots:** isolated plumes of tectonic activity known as **hot spots** (where a plume of magma punches a hole through the lithosphere and crust and erupts at the surface) may give rise to seismic activity.

Earthquake magnitude and frequency

The magnitude of earthquakes is measured by the:
+ **Richter scale:** a logarithmic scale from 1–10, so an earthquake of 7 is 10 times greater than one measuring 6
+ **moment magnitude scale (MMS):** this identifies the energy release on a scale of 1–10
+ **Mercalli scale:** this measures earthquake intensity through the impact on people and structures. The scale is 1–12 – 1 is detected only by seismographs, 12 causes total destruction.

Large earthquakes (> 8.0 on the Richter scale in magnitude) have struck the Earth at a record high rate since 2004; however, the increased frequency has not been statistically different from what one might expect to see by random chance.

Impacts of seismic events

Earthquakes of a high magnitude can devastate large areas and kill or injure tens of thousands of people. Collapsed buildings are the main cause of death, but in the aftermath of an earthquake, fire, disease and damage to infrastructure add to the suffering and death toll.

Clearly the size of the event is the main determinant of impact, but population density, degree of preparedness, time of day and level of economic development can also affect the impact.

Table 5.4 summarises some of the primary and secondary hazards and impacts.

Table 5.4 Earthquake hazards and impacts

Hazards	General impacts
Primary hazards: + ground shaking + ground rupture Secondary hazards: + landslides + liquefaction + tsunamis + rockfalls + debris flows	+ Loss of life + Injury + Destruction of buildings + Destruction and disruption to infrastructure – water, sewerage, transport, and utilities such as gas and electricity + Fires + Release of hazardous materials + Spread of disease + Potential of floods if dams collapse + Disruption to the economic functioning of affected locations + Looting + Food security issues + Erosion and destruction of natural habitats + Psychological trauma

Specific hazards and impacts can be studied through a named example (as in the revision activity below, based on a recent example studied in class).

Revision activity

In class you will have studied an example of a recent seismic event and you should create revision notes about it. The focus should be on the impacts of the event and the human responses. You will also need a brief description of the event – location, date, sequence of events. A table can be used to summarise the impacts and human response. Remember to categorise the impacts as environmental, social, economic and political, and also short or long term. You may also colour code impacts as primary and secondary.

Now test yourself

22 Explain the link between destructive plate margins and a high concentration of earthquakes.

23 Why do earthquakes sometimes occur away from plate margins?

Answers on p. 264

TESTED ○

24 What are the primary and secondary hazards generated by earthquakes?

25 List the factors that influence the human impact of earthquakes, giving your explanation for each.

Answers on p. 264

Short- and long-term responses

Prediction

Prediction is very difficult. Regions at risk can be identified. Attempts to predict include:

+ monitoring groundwater levels
+ monitoring release of radon gas
+ strange animal behaviour
+ measuring magnetic fields.

Hazard zone maps can be acted upon by national and local planners.

In terms of **predictability**, observations of the following can help:

+ crustal movement
+ changes in electrical conductivity
+ strange and unusual animal behaviour
+ historic evidence.

> **Exam tip**
>
> Rich countries, such as Japan and the USA, are much more able to invest in the most advanced protection. In poorer parts of the world, building collapse leads to huge loss of life and challenges for the emergency services. Often expertise is sent in from advanced economies. However, poorer countries which are affected by hazards or by a high frequency of hazard events often build up specialist knowledge.
>
> These points are useful in answering questions on evaluating the ability of countries at different levels of economic development to respond to hazard events.

Prevention

Prevention of seismic natural events is impossible. Some scientists are looking into reducing the friction caused at conservative plate boundaries.

Protection/preparedness and mitigation

Making sure that the population of areas at risk is well informed regarding protection and safety during an earthquake is essential. Measures include:

+ promoting understanding of earthquakes and their effects
+ understanding how homes can be adapted to better withstand the impacts of earthquakes
+ ensuring safety drills are well practised and understood
+ mitigation of impact by appropriate insurances
+ building of hazard-resistant structures, for example rubber shock absorbers in foundations and cross-bracing to hold structures together and allow flexibility as a building shakes
+ fire prevention – 'smart meters' have been developed to cut off gas supplies if an earthquake of a high magnitude occurs
+ careful education, training and preparation of emergency services
+ land-use planning – constructing certain types of building in low-risk areas
+ tsunami protection, for example automated warning systems; however, large tsunamis can overwhelm these
+ securing heavy items within the home and moving breakable items to low-level storage
+ developing early-warning systems.

> **Exam tip**
>
> When analysing protection and prevention methods, remember that tsunamis present an exceptional hazard as effective protection is extremely difficult and the impact is often a long way from its origin.

Check your understanding and progress at **www.hoddereducation.co.uk/myrevisionnotesdownloads**

Adaptation

Adaptation of the environment depends on the level of economic development of a country, education, and national and regional priorities. Measures may include:

+ land-use planning – using open spaces as safe areas, building high-risk buildings, e.g. hospitals, in low-risk areas
+ providing emergency services with specialist equipment.

Now test yourself

TESTED ⬤

26 Outline four responses to mitigate the impact of earthquakes.

Answer on p. 264

> **Exam tip**
>
> Choosing a volcanic and seismic event from contrasting geographical settings allows you more flexibility in your answers to examination questions and aids your understanding.

A recent seismic event – impacts and human response

> **Exam tip**
>
> It is useful to be able to compare the responses to seismic events between countries of differing levels of economic development – for example, volcanic eruptions in Japan and Indonesia or earthquakes in the USA and Nepal. This is useful when asked to 'evaluate' responses.

> **Revision activity**
>
> For a named advanced country and a low-income developing country, make your own revision notes on the positive aspects of their response to a seismic event and the challenges they face. Remember that low-income countries can build up a lot of expertise in dealing with hazard events.

> **Exam tip**
>
> For an explanation of seismic and volcanic events you must be clear on processes. For volcanoes, be clear on the products of the eruption.

Storm hazards

REVISED ⬤

The nature of tropical storms and their underlying causes

Tropical storms are intense, low-pressure systems that develop in the tropics. They are referred to as hurricanes (in the Atlantic), cyclones (generally in southern Asia) and typhoons (in the western Pacific).

> **Exam tip**
>
> When explaining the origin and underlying causes of storm events, remember that the relationship between oceanic temperatures and world climate is complex and year to year there is considerable uncertainty.

The storms originate from an area of low pressure, where surface heating results in warm air being drawn in in a spiralling manner. Several conditions need to be present:

+ an oceanic location with sea temperatures over 27°C
+ an ocean depth of more than 70 m (the moisture provides latent heat and the cold water is not stirred up by the storm from the deeper ocean)
+ a location 5° north or south of the equator – so that the Coriolis force brings about maximum rotation
+ convergence of air in the lower atmosphere
+ rapid outflow of air in the upper atmosphere.

111

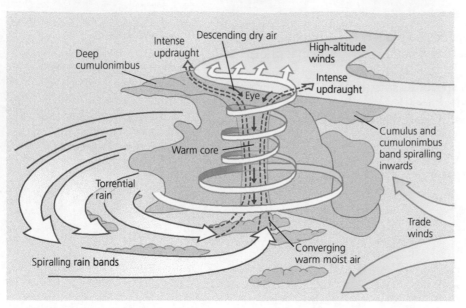

Figure 5.13 Structure of a tropical storm

Figure 5.13 shows the structure of a tropical storm.

The sequence of formation is as follows:
1 Warm ocean water causes a large amount of water evaporation.
2 Winds converge close to the ocean surface, forcing air upwards.
3 The air is unstable and winds rise rapidly.
4 Warm rising air condenses to form cloud and rain. The heat generated from condensation warms the surrounding air and it rises, forming an intense 'up draught'.
5 Dry, cooler air from the upper atmosphere descends.

Spatial distribution

Tropical storms are distributed between 5° and 20° north and south of the equator. The global distribution is shown in Figure 5.14.

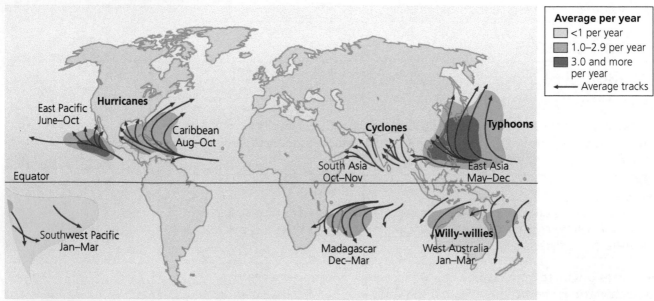

Figure 5.14 Global distribution and seasons of tropical storms

Now test yourself

TESTED ⚪

27 State three conditions necessary for a tropical storm event.

28 Give a sequenced explanation of the formation of a tropical storm.

Answers on p. 264

Check your understanding and progress at **www.hoddereducation.co.uk/myrevisionnotesdownloads**

Magnitude, frequency and predictability

The main form of measurement is the Saffir–Simpson scale, a scale of 1–5 based on pressure at the centre, wind speed, storm surge and damage potential. As an example:

+ 1 = minimal damage, >980 mb pressure, 119–153 winds km/h^{-1}, 1.2–1.5 m storm surge
+ 5 = catastrophic damage, <920 mb pressure, >249 wind speed km/h^{-1}, >5.5 m storm surge.

Every year around 80–100 tropical storms develop around the world.

They are extremely difficult to predict as their onset is rapid and although they may move slowly at first, their path is erratic and therefore difficult to predict.

Impacts of storm hazards

Impacts depend on:

+ storm intensity
+ speed of movement
+ distance from the sea
+ preparation of the community
+ warnings and community response.

Types of impact include:

+ coastal flooding
+ river flooding
+ storm surges
+ damage to roofing or complete roof structure failure
+ structural damage on a scale relating to the Saffir–Simpson scale – window and door damage; some small structures and mobile homes blown away
+ varying degrees of flooding to homes
+ damaged vehicles
+ uprooted trees
+ roads blocked by debris and fallen trees
+ power lines damaged due to high winds
+ destruction of crops, leading to food security issues
+ landslides and mudslides due to intense rainfall.

Storm surge A rapid rise in sea level in which high winds (in a tropical storm) push the sea upwards and in the direction of the coastline.

Food security Access to safe and nutritious food in sufficient quantities for individuals to lead a healthy life.

Revision activity

For a named storm event, make notes about the impacts of storm hazards. Try to link your notes to the types of impact listed under the bullets above. Add information on specific economic, physical and human impacts.

Responses and management

Preparedness

+ The National Hurricane Center in Florida, USA can access data from geostationary satellites. Using weather aircraft, the USA also maintains round-the-clock surveillance of tropical storms that have the potential to become hurricanes.
+ Weather satellites can be used to track the location, path, size and strength of hurricanes.
+ Cyclones have an erratic path and it is therefore often difficult to give warnings. In poorer areas where communication is limited, there is less warning.
+ Preparedness programmes, such as the Cyclone Preparedness Programme in Bangladesh, can help with preparation.

Making links

It is useful to compare responses and management to the three other types of hazard event within this unit. See responses and management to:

+ volcanic hazards (page 108)
+ earthquake hazards (pages 111–12)
+ wildfires (page 117).

Prevention

As with all natural hazards, tropical storms cannot be prevented. Research is looking into the potential to force cyclones to release more water over the sea, thereby weakening the system over the land.

Prevention techniques considered in the last 80 years include:
+ cooling oceans with ice to reduce evaporation
+ blowing black soot into the storm to change the radiation balance.

Protection and mitigation

Hurricane/cyclone drills can be practised. People can be evacuated, but homes and businesses then need to be protected once an area has been evacuated. People can strengthen their homes to withstand high winds.

Land-use planning can identify areas most at risk and ensure that certain building and developments do not take place there. Sea walls, breakwaters and flood barriers can be used to protect coastal areas.

The level of damage can be reduced by making roads, bridges and public buildings more resilient.

Adaptation

This may include:
+ land-use planning
+ building seawalls, breakwaters and flood barriers
+ retrofitting buildings to make them wind resistant.

In richer areas people can afford to take out insurance against damage. In poorer areas there must be an understanding of the aid available that can be put in place quickly.

Two recent storms in contrasting areas of the world – impacts and human responses

> ### Revision activity
>
> In class you will have studied two recent tropical storms in contrasting areas of the world and you should create revision notes on them. The focus should be on the impacts of the events and the human responses. Write a brief description of each storm – location, date, sequence of events. Create a table to summarise the impacts and human response. Remember to categorise the impacts as environmental, social, economic and political, and that for these examples there must be contrasts in the impacts and level of response.

> ### Now test yourself
>
> 29 Why are tropical storms difficult to predict?
> 30 How can people be protected against tropical storms?
> 31 State three factors that will determine the impact of a tropical storm.
> 32 Name three natural hazards caused by tropical storms.
>
> **Answers on p. 264**
>
> TESTED ○

> ### Exam tip
>
> Any assessment of impact must show a balanced understanding of the:
> + magnitude and nature of the hazard
> + people at risk
> + ability of the country to respond and mitigate hazard impact.

Fires in nature

REVISED ○

Nature of wildfires

Wildfires are a natural process in many ecosystems and can bring about benefits:
+ A regular occurrence of small fires can reduce the amount of fuel build-up, thereby lowering the likelihood of a large and more dangerous fire.
+ Fires often remove alien plants that compete with native species for nutrients and space.
+ The ashes that remain after a fire add nutrients (often locked in older vegetation) to the soil.
+ Fires can also provide a way of controlling insect pests by killing off the older or diseased trees and leaving the younger, healthier trees.

Fires on a large scale are major events which cause widespread destruction and kill wildlife. Ground temperatures can rise above 1000°C and at higher levels fire can spread rapidly through the canopy.

The exact nature of the fire depends on the plants involved (savannah grassland can ignite quickly), strength of winds (high winds cause greater risk), levels of humidity and the behaviour of the fire itself.

Check your understanding and progress at **www.hoddereducation.co.uk/myrevisionnotesdownloads**

Causes of fires: natural and human

Natural causes

+ Lightning is the major natural cause of fires.
+ Climate will affect the frequency of electrical storms – low rainfall and hot days will lead to convectional storms.
+ Prolonged dry conditions also lead to drought and increase the risk of wildfires.

Human causes

Increasingly, human impact is the cause of fires:

+ falling power lines
+ lack of consideration for the effects of camp fires
+ discarded cigarettes
+ arson.

In the tropical rainforest areas (the Amazon Basin and Indonesia) some fires started as a way of clearing the forest have quickly got out of hand and spread.

Impacts: primary/secondary, environmental, social, economic, political

The countries mainly impacted by wildfires are Australia, France, Italy, Turkey, Spain, Greece, Portugal, the USA (the states of California and Florida in particular) and Canada. Mainly rural areas are affected but increasingly the rural–urban fringe is also vulnerable.

Table 5.5 outlines the various impacts of fire hazards. See the accompanying revision activity for completing the table.

Table 5.5 Impacts of fire hazards

Impact	Primary or secondary	Environmental, social, economic or political
Loss of crops, timber and livestock		
Loss of life		
Loss of property		
Release of pollutants, e.g. toxic gases and particulates		
Loss of wildlife		
Damage to soil structure and nutrient content		
Organic matter protects the soil from erosion – once this is reduced, the soil is exposed		
Soil can become more resistant to rainfall, overland flow will increase and infiltration decrease		
Evacuation and consequent need for shelter, food and resources		
Increased flood risk due to less interception and increased soil erosion and silting of rivers		

Revision activity

Copy and complete Table 5.5 by categorising the impacts into primary and secondary and indicating whether each impact is environmental, social, economic and political. Give a brief explanation for your choice (some impacts may fit more than one category).

Now test yourself

TESTED

35 Name two low-cost and two high-cost attempts to mitigate the impact of wildfires.
36 How do wildfires lead to long-term human suffering?

Answers on p. 264

Exam tip

Be aware that natural hazards are often interconnected – droughts, wildfires and famine are often linked.

Now test yourself

33 How can wildfires be beneficial?
34 Under what circumstances do humans start fires?

Answers on p. 264

TESTED

Responses: preparedness, mitigation, prevention

In the first instance, responding to wildfires means extinguishing the fire hazard. This in itself is dangerous work. Once this has been achieved, governments and communities must address the impact and prepare for future possible events.

Figure 5.15 outlines the various strategies for preparing for fires, as well as strategies for mitigation and prevention.

Figure 5.15 Fire hazard preparedness, mitigation, prevention and adaptation

Impacts

Primary effects
+ Loss of crops, timber and livestock.
+ Loss of life.
+ Loss of property.
+ Release of toxic gases and particulates.
+ Loss of wildlife.
+ Damage to soil structure and nutrient content.

Secondary effects
+ Evacuation.
+ Increased flood risks.

> **Revision activity**
>
> In class you will have studied a recent wildfire event and you should make revision notes about it. The focus should be on the impacts of the event and the human responses. You will also need a brief description of the event – location, date, sequence of events. Create a table or star diagram summarising the impacts and human response. Remember to categorise the impacts as environmental, social, economic and political.

> **Revision activity**
>
> Make revision notes on the local-scale case study covered in class. Make sure that you focus on the physical nature of the hazard and how the character of the local community reflects the presence, impact and response to the risk.

Check your understanding and progress at **www.hoddereducation.co.uk/myrevisionnotesdownloads**

Case studies

A multi-hazardous environment beyond the UK, for example the Philippines

This case study should:

+ illustrate and analyse the nature of the hazards and the social, economic and environmental risks presented
+ investigate how human qualities and responses such as resilience, adaptation and mitigation, and management contribute to its continuing human occupation.

What do I need to know?	Content and suggested revision methods
The nature of the hazards	Causes:
	There is a range of hazards affecting the Philippines due to its location on the western rim of the Pacific Ring of Fire and being in the path of tropical storms tracking from the Pacific towards South East Asia.
	Hazards:
	Primary hazards:
	+ earthquakes, e.g. off Mindanao, 1976
	+ volcanic activity, e.g. Pinatubo, 1991
	+ tropical storms, e.g. Typhoon Mangkhut, 2018.
	Secondary hazards:
	+ tsunamis, fires, landslides (from earthquakes)
	+ lahars (from volcanic eruptions)
	+ flooding and landslides (from tropical storms).
The social, economic, and environmental risks presented	Risks:
	+ **Social:** poorly constructed housing is vulnerable to destruction by hazards.
	+ **Economic:** looting, poverty.
	+ **Environmental:** deforestation leading to unprotected slopes where landslides easily occur.
	Create a colour-coded diagram, as shown below, summarising the causes, hazard events and hazard risks affecting the Philippines.
Continuing human occupation due to management, resilience, adaptation and mitigation	Notes on:
	+ **mitigation** – land-use planning, construction, preventative measures
	+ **resilience and adaptation** – post-disaster relief, forecasting, evacuation, fatalistic acceptance
	+ **management** – planning by central government for long-term solutions and the involvement of NGOs (for example the Red Cross).
	Practise answering an exam question on why people continue to live in multi-hazard environments. Your answer should look at management, mitigation, resilience and adaptation and can form part of your revision notes.

A local-scale specific place in a hazardous setting, for example Gili Trawangan, Indonesia

This case study should:

+ illustrate the physical nature of the hazard
+ analyse how the economic, social and political character of its community reflects the presence and impacts of the hazard and the community's response to the risk.

What do I need to know?	Content and suggested revision methods
The physical nature of the hazard	A bulleted/sequential summary of the cause of earthquakes in this location or an annotated diagram would be a good way to organise the content. For example: ✦ Gili Trawangan is part of an island arc structure where → the Indo-Australian Plate is moving north, subducting the south-east moving Eurasian Plate. ✦ Friction occurs. ✦ The subducting plate gets stuck. ✦ When the stress is released and the plate becomes 'unstuck', energy is released as an earthquake.
How the economic, social and political character of the community reflects the: ✦ vulnerability to the hazard ✦ impact of the hazard ✦ response to the risk of the hazard.	1 A table is a good way to organise your revision notes. Make a column each for **vulnerability**, **impacts** and **responses**. Highlight points as either **economic**, **social** or **political**. For example, having structurally unsound buildings is a **social vulnerability**. 2 Alternatively, as there will be overlap, you could use a Venn diagram to show the impacts of the hazard. For example, looting and late arrival of the police to deal with the hazard is both social and economic.

Exam practice

1 Using the data in Table 5.6 below, analyse the variations in casualties. [6]

2 Explain the structure and formation of a tropical storm. [6]

3 Assess the effectiveness of a pre-disaster management plan for a located seismic event. [9]

4 With reference to a specific example, assess the attempts made to manage the impacts of a recent wildfire event. [9]

5 Explain the features of a destructive plate margin. [9]

Table 5.6 Volcanic eruptions and casualties

Volcano	Volcanic explosivity index (VEI)	Year	Approximate number of casualties
Mount Pelee, Martinique	4	1902	33,000
Santa Maria, Guatemala	6	1902	6,000
La Soufriere, St Vincent and the Grenadines	4	1902	1,680
Kelud, Indonesia	4	1919	5,000
Mount Lamington, Papua New Guinea	4	1951	3,000
Mount Agung, Indonesia	5	1963	1,600
Mount St Helens, USA	5	1980	60
El Chichown, Mexico	5	1982	3,500
Nevado del Ruiz, Colombia	3	1985	23,000
Pinatubo, Philippines	6	1991	850

Answers and quick quizzes online

Exam skills

Opportunities to practise geographical skills within this topic include:

+ statistical analysis of hazard-event data – data on magnitude, casualties, timescale
+ application of calculations such as interquartile range, standard deviation
+ measures of correlation, such as scatter graphs and Spearman's rank correlation coefficient
+ analysis of specific graphs and histograms showing, for example, hazard frequency, magnitude or impact
+ interpretation of geospatial data (for example geological maps, hazard maps)
+ interpretation of maps (for example showing location of hazard events or isoline maps showing numbers of casualties) and photographs
+ analysis of patterns of distribution on maps.

Summary

+ Extreme natural events such as earthquakes and wildfires become natural hazards when they adversely affect people.
+ Human responses to hazards can be explained through models, such as the Park model and the hazard management cycle.
+ Be clear on description of the spatial distribution of hazards and the need for clear explanation using subject-specific vocabulary to explain the pattern.
+ You should have a detailed understanding of the processes and landforms associated with plate tectonics, vulcanicity, seismicity, tropical storms and wildfires.
+ Well-annotated sketch diagrams should be practised to aid complex explanations.

+ The impacts of natural hazards can be categorised into primary and secondary. It is important to be clear on these for all types of hazard in this unit.
+ Impacts can also be social, economic, political and environmental, as well as short and long term.
+ Understand different mitigation, protection and prevention methods for each hazard and the factors that affect the ability of different countries to put these measures into practice.
+ In case studies and examples, avoid generalised statements – the complexity and individual nature of located events should be understood.

119

Ecosystems and sustainability

The concept of biodiversity

Biodiversity is the measure of the number, variety and variability of living organisms.

It includes diversity within species, between species and among ecosystems. The concept also covers how this diversity changes from one location to another and over time.

> **Ecosystem** A system in which organisms interact with each other and with the environment.

Biodiversity is a part of the **global ecosystem**, which covers all living things and the environment in which they live. The link between biodiversity and the functioning of ecosystems is that biodiversity supports healthy ecosystems and ecosystems provide the basis for life support.

Measures of biodiversity:

+ **Indicator species** (when certain species are used as an indication of environmental conditions) is often presented as evidence for planning and conservation.
+ **Species richness** is a more commonly used measure as it provides a direct link between the number of species and biodiversity.
+ The **Living Planet Index (LPI)** presents trends in vertebrate species in order to determine the 'health' of ecosystems.

> **Exam tip**
>
> It is important to be able to relate the concept of biodiversity to local and global examples.

> **Now test yourself**
>
> TESTED
>
> 1 What is the link between biodiversity and global ecosystems?
> 2 What are indicator species?
>
> **Answers on p. 264**

Local and global trends in biodiversity

Local trends

The Biological Records Centre (BRC), part of the Centre for Ecology & Hydrology (CEH), provides a focus for the collation and management of species observations (biological records).

The BRC works on a local basis across the UK on more than 80 recording schemes.

The recording leads to publication of atlases, data and other online resources and therefore provides essential information for research and conservation.

Table 6.1 shows the link between the potential causes of biodiversity change and the impacts of this on a specific ecosystem – a blanket bog.

Revision activity

Make brief notes on a biodiversity change in a local example studied in class. This may be based on freshwater species or habitat loss due to agriculture or urban expansion. An example is shown in Table 6.1.

Table 6.1 Causes and impacts of biodiversity change on a blanket bog on Kinder Scout, Derbyshire

Cause	Consequence	Impacts and potential impacts
Climate change		
Increased mean temperatures	Longer growing season	+ Bracken has become invasive at higher altitudes + Potentially increased nitrogen deposition, leading to changes in plant species + Mire vegetation (that is, plant types that thrive in wet conditions) may become less dominant
Hotter summers	Increased evapotranspiration and lowering of water table	+ Changes in species composition + Increase in the release of dissolved organic carbon, leading to declining water quality
Drier summers	Drought Drier ground conditions Wildfire	+ Shift in the dominance of species + Possible wind erosion of dried peat + Possible increased agricultural potential + Changes in red grouse populations
Wetter winters	Increased overland flow	+ Loss in peat stability; increased slides
Storm events	Increased rainfall intensity	+ Gullying erosion
Other influences		
A long period of grazing	Over-grazing	+ Loss of biodiversity
Access to walkers from nearby industrial towns and cities since the 1940s	Footpath erosion Damage to surface vegetation that holds the peat in place	+ Large areas of peat lost + Exposed dried-out peat subject to wind and gullying erosion
Increased wildfires	Loss of peat	+ Reduction in species numbers
Draining of peat bog	Lowering of water table	+ Sphagnum mosses have been replaced by other species such as heather and moor grass + Species of moorland birds that nest and feed in the heather, such as golden plover and curlew, as well as the mountain hare, are threatened
Industrial pollution		+ Reduction in number of species

Global trends

Global biodiversity has declined steadily. Virtually all of the Earth's ecosystems have now been dramatically transformed through human actions. Scientists believe that we are currently in a sixth period of mass extinction, this time caused by humans.

Data indicating declining biodiversity include:
+ Populations of vertebrate species fell by 52 per cent between 1970 and 2010.
+ There was a 56 per cent reduction of biodiversity in the tropics between 1970 and 2010.
+ The biomes with the highest rates of degradation in the last half of the twentieth century were temperate, tropical, flooded grasslands and tropical dry forests.
+ The LPI for freshwater species showed an average decline of 76 per cent between 1970 and 2010.
+ Marine species declined by 39 per cent between 1970 and 2010, the greatest decline being in the tropics and the Southern Ocean.

It is important to understand the:
+ causes of declining biodiversity
+ impacts of declining biodiversity.

Making links

For more on the causes and impacts of climate change see Chapter 1.

Biome A large-scale ecosystem, usually at a continental scale, based on distinct plant and animal species which depend on particular climatic patterns.

The main causes of declining biodiversity are summarised in Figure 6.1. They include:

+ exploitation through hunting and fishing
+ habitat degradation/change and habitat loss, particularly from agriculture
+ urban development
+ energy production.

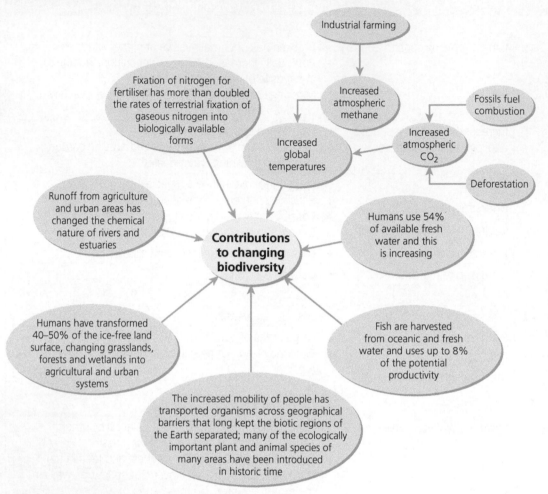

Figure 6.1 Contributions to changing biodiversity

Climate change is expected to exacerbate risks of extinctions of both animals and plants, floods, droughts, population declines and disease outbreaks.

Recent changes in climate have already had significant impacts on biodiversity and ecosystems. As climate change becomes more severe, the harmful impacts on ecosystems are expected to increase.

Biodiversity is a key factor determining human well-being. Biodiversity loss has direct and indirect negative impacts on several factors:

+ **Food security:** some communities depend on indigenous plants; also, insect species are an important natural pest control and crop pollinators.
+ **Vulnerability to some natural disasters:** for example, the loss of mangroves and coral reefs, which are natural buffers against floods and storms.
+ **Health:** a balanced diet depends on the availability of a wide variety of foods, which in turn depends on the conservation of biodiversity.
+ **Clean water:** the continued loss of forests and the destruction of watersheds reduce the quality and availability of natural water supplies.
+ **Basic materials:** biodiversity provides basic plant and animal resources needed in a variety of sectors, such as agriculture, 'ecotourism', pharmaceuticals, cosmetics and fisheries.

Now test yourself

3 Outline three reasons why global biodiversity is declining.

4 How has the increased mobility of people affected biodiversity?

5 How does a decline in biodiversity impact food security?

Answers on p. 264

TESTED

Check your understanding and progress at **www.hoddereducation.co.uk/myrevisionnotesdownloads**

Ecosystems and their importance for human populations

Humans are an integral part of ecosystems. Ecosystem services are the benefits that people obtain from ecosystems. The **Millennium Ecosystem Assessment** (MEA) analysed 24 ecosystem services and found that 15 were being degraded or used unsustainably.

There are four categories of ecosystem service:
1 **Provision services** (food, water).
2 **Regulating services** (climate, water storage and purification).
3 **Supporting services** (soils and nutrients).
4 **Cultural services** (recreation and spiritual).

Changes in biodiversity and ecosystem services are due to:
+ natural causes
+ demographics
+ economics
+ socio-political factors
+ culture and religion
+ science and technology.

Current changes are predominantly driven by human factors resulting from growing populations and increased consumption per capita.

> **Ecosystem services** The benefits people obtain from ecosystems.

> **Now test yourself**
>
> 6 How can loss of forests affect water supply?
> 7 Explain how certain demographics, such as growing population, can affect ecosystem services.
>
> **Answers on p. 264**
>
> TESTED ○

Human populations in ecosystem development and sustainability

Population growth and climate change are the two main determinants of future sustainability of ecosystem development and biodiversity change.

There are two contrasting responses (a positive and a negative feedback loop), which are outlined in Figure 6.2.

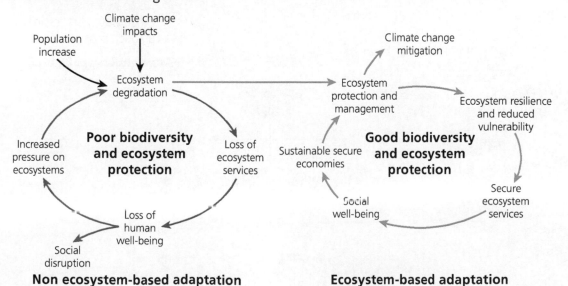

Non ecosystem-based adaptation

Ecosystem-based adaptation

Figure 6.2 Beating the vicious cycle of poverty, ecosystem degradation and climate change

123

Revision activity

Using Figure 6.2 and the concept of positive and negative feedback, construct two flow diagrams to summarise how a) population growth and b) economic development can change ecosystem services.

Ecosystems and processes

REVISED

Nature of ecosystems

- An ecosystem is a community of living (**biotic**) and non-living (**abiotic**) things that work together and interact.
- The physical environment provides the energy, living space and nutrients that plants and animals need to survive.
- They can exist at any scale – biome to puddle.
- All parts of the ecosystem work in balance.
- They are open systems as energy and materials cross ecosystem boundaries (see Figure 6.3).

Exam tip

The key to understanding the characteristics and behaviour of ecosystems is the relationship between the physical environment and the adaptation of living organisms.

Figure 6.3 Inputs and outputs in an open ecosystem

Structure

As shown in Figure 6.4, ecosystems have two components:

- biotic factors
- abiotic factors.

Now test yourself

8 Differentiate between biotic and abiotic components of ecosystems and give two examples of each.

9 Name two transfer processes operating in an open ecosystem.

Answers on pp. 264–65

TESTED

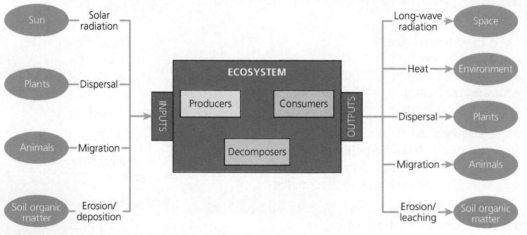

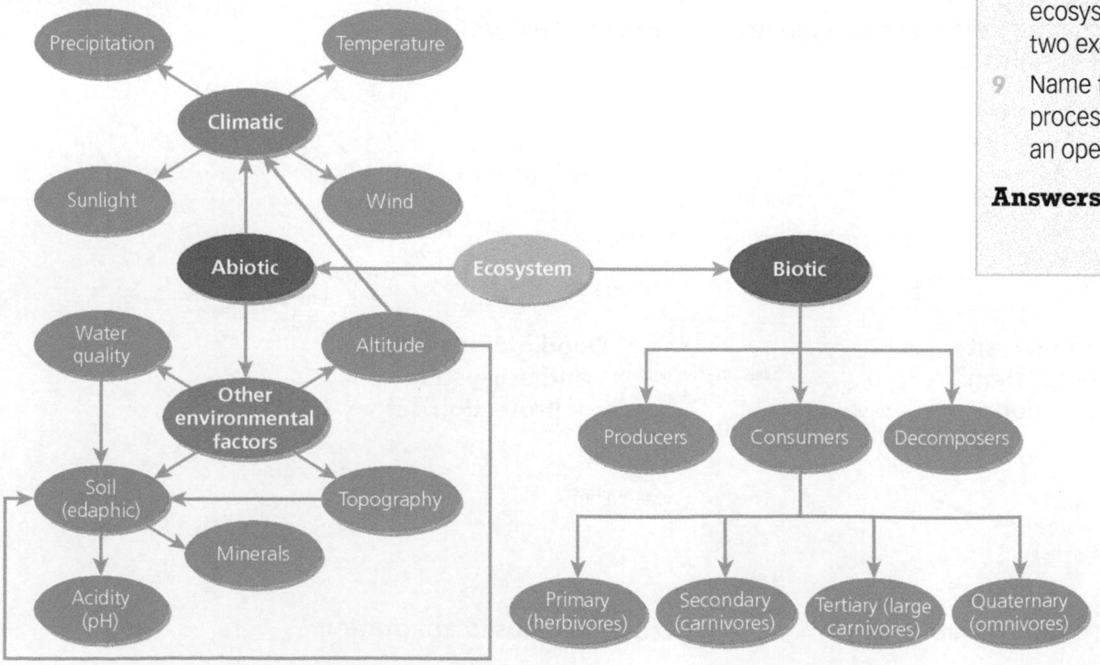

Figure 6.4 Biotic and abiotic components in an ecosystem

Biotic components

Biotic components are classified into three further groups:

+ **Producers** – green plants capable of converting solar energy into chemical energy by the process of photosynthesis. They are known as autotrophs as they manufacture their own food.
+ **Consumers** – known as heterotrophs as they do not manufacture their own food but depend on producers. There are four groups of consumers:
    + **primary consumers** (herbivores that feed on producers)
    + **secondary consumers** (primary carnivores that feed on herbivores)
    + **tertiary consumers** (large carnivores that feed on secondary consumers)
    + **quaternary consumers** (omnivores – large carnivores that feed on tertiary consumers).
+ **Decomposers** – bacteria and fungi, which break down dead organic matter of producers and consumers and release it into the environment.

Abiotic components

Abiotic factors include:

+ climate
+ soils
+ topography
+ altitude.

> **Exam tip**
>
> Be clear on the interconnections between biotic and abiotic components. This is also the foundation of understanding impacts and responses to changes in ecosystems.

> **Revision activity**
>
> Figure 6.4 provides a generic representation of the biotic and abiotic components of an ecosystem. Produce a copy of this diagram with specific information and examples relating to the two following biomes:
>
> a) tropical rainforest
>
> b) savanna grassland.

> **Making links**
>
> Further information on tropical rainforest biomes and savanna grassland biomes can be found on pages 134–35.

Energy flows, trophic levels, food chains and food webs

Figure 6.5 shows the generalised energy flows through an ecosystem.

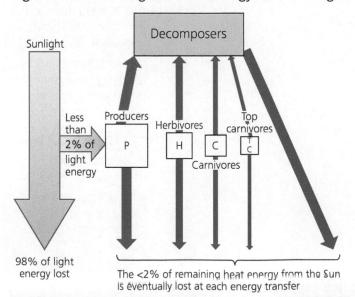

Figure 6.5 Generalised energy flow and heat loss through an ecosystem

Sunlight is captured by the leaves of green plants and this is the **primary energy** source of most ecosystems. This energy is then transferred between producers and consumers via food chains (Figure 6.6) or more complex food webs (Figure 6.7).

> **Making links**
>
> See the carbon cycle in Chapter 1 page 24.

> **Food chain** The sequence of transfer of nutrients and energy from one organism to the next in the order in which they eat one another.
>
> **Food web** A matrix of feeding relationships that resembles a web.

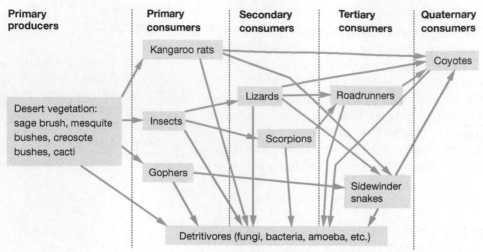

Figure 6.6 The food chain of an oak woodland

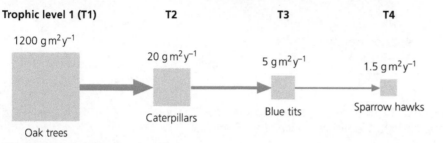

Figure 6.7 A desert food web

The flow of energy occurs in a number of stages called **trophic levels** (see Figure 6.8), which relate to the levels of different consumers as above:

+ T1: autotrophs
+ T2: primary consumers
+ T3: secondary consumers
+ Tn: a top predator – depending on the length of the food chain it may be a tertiary or a quaternary consumer.

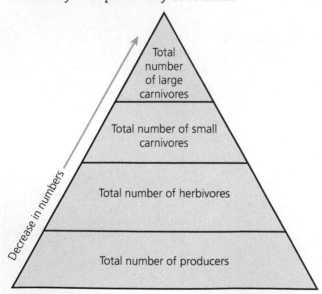

Figure 6.8 A number pyramid showing trophic levels

At each trophic level energy is lost. This is because organisms convert only a small percentage of the energy they consume into living tissue. Energy is used to keep the organism alive and energy is lost as heat in respiration. As a result:

+ the number of trophic levels is limited
+ the biomass declines at each trophic level
+ there is a reduction in the number of organisms at each trophic level.

The interconnection of food chains forms food webs.

Check your understanding and progress at **www.hoddereducation.co.uk/myrevisionnotesdownloads**

The application of systems concepts to ecosystems

Figure 6.3 (page 125) shows how ecosystems are viewed as systems.

+ The **inputs** include sunlight, soil, plants and animals.
+ The **outputs** include heat and waste products.
+ **Flows** and **processes** within the system include photosynthesis and respiration.
+ **Stores** include trees.

The system is in **dynamic equilibrium**, but disturbance can bring **positive** and **negative feedback** – for example, plentiful food supply leads to increased numbers through the trophic levels.

Concept of biomass and net primary production

Biomass is the mass of living organisms (plants, algae, bacteria) in a particular ecosystem. It is usually expressed as the average dry mass per unit area, for example tha^{-1} or kgm^2.

Gross primary productivity (GPP) is the total energy fixed by plants in a community through photosynthesis.

Net primary productivity (NPP) subtracts the proportion of energy in GPP used by plants for respiration.

Succession – seral stages, climatic climax, sub-climax and plagioclimax

Ecological succession is the sequence of vegetation changes through time.

+ **Primary succession** is the change on a site previously uncolonised – for example, with bare rock, the process starts with **pioneer species**.
+ **Secondary succession** occurs on sites that have been vegetated but where the vegetation cover has been destroyed, perhaps by fire, for example.

As succession takes place there are key features of each stage:

+ more complex structure
+ more biomass
+ more species
+ greater NPP
+ greater flows of energy and nutrients.

Each vegetation stage is referred to as a **sere**, which modifies the environment, allowing new species to grow.

Climax vegetation is when the vegetation is in balance with the natural environment – it may persist indefinitely.

+ **Climatic climax:** when climate controls the climax vegetation, for example the tropical rainforest.
+ **Sub-climax:** where other factors, such as slope or soils, dominate.
+ **Plagioclimax:** where human activities (such as burning, planting, grazing and draining) control the climax.

Table 6.2 summarises some of the arresting factors in ecological succession.

Table 6.2 Some arresting factors in ecological succession

Arresting factor	Possible cause(s)
Topoclimax	A change in topography. This could be a landslide, a volcanic eruption (lava or ash) or deposition of mud following a river flood.
Hydroclimax	A change in drainage. This could be caused by a raised water table following increased precipitation.
Biotic climax	The introduction of an alien species to an area. For example, rabbits are the most significant known factor in species loss in Australia.
Plagioclimax	An environment maintained by management. This could be forest clearance in Amazonia for ranching, for example. It includes all agriculture, sports fields, parks, gardens, etc.

> **Revision activity**
>
> Produce a summary flow diagram that explains **succession**.
>
> Your diagram must show the links between climatic climax, sub-climax and plagioclimax vegetation and include key terms such as secondary and primary succession, seres and pioneer species development.

Succession starts on a variety of surfaces and their names reflect this:

+ **lithosere** – exposed rock
+ **psammosere** – bare sand
+ **hydrosere** – fresh water
+ **halosere** – salt water.

Making links

The concepts of a hydrosere and lithosere are covered later in this chapter (pages 135–36) in the context of the British Isles.

Figure 6.9 summarises some important general points about succession.

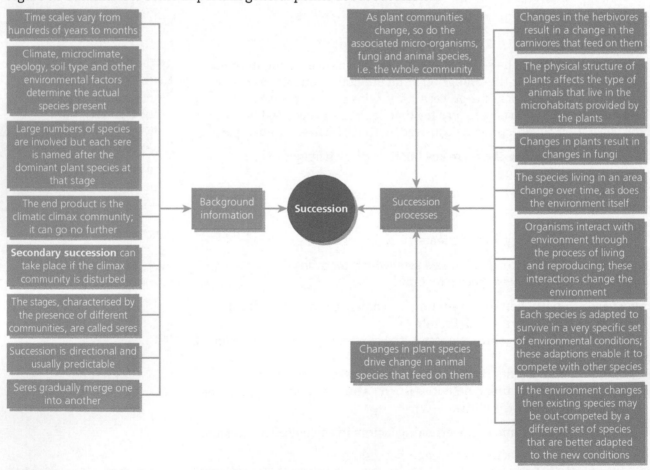

Figure 6.9 Succession: important points

Mineral nutrient cycling

Nutrients needed for plant growth (phosphorus, nitrogen and iron) are recycled through the ecosystem between biomass, litter and soil stores.

+ Inputs to the cycle include rock weathering and precipitation.
+ Outputs include surface runoff and leaching.
+ Nutrient stores are in soil, litter (dead organic matter on top of the soil) and biomass.

Check your understanding and progress at **www.hoddereducation.co.uk/myrevisionnotesdownloads**

Figure 6.10 shows the Gersmehl diagram, which summarises a mineral nutrient cycle. The size of the stores (circles) and flows (arrows) can be adapted to different environments to reflect the volume of nutrients being transferred.

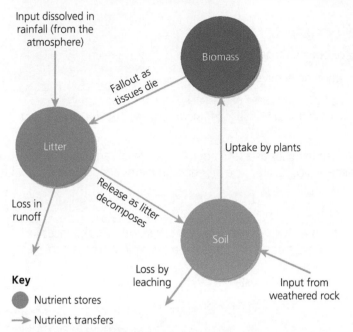

Key

● Nutrient stores

→ Nutrient transfers

Figure 6.10 A model of a mineral nutrient cycle

Nature of terrestrial ecosystems

Ecosystem refers to the interaction of organisms with each other and with their environment. The relationships and interactions within this form an ecological unit.

Terrestrial ecosystems exist on land. They range in scale from local to global (biomes). Table 6.3 gives an example of a terrestrial ecosystem – the chalk downlands of the South Downs.

Table 6.3 An example of a chalk downland terrestrial ecosystem – the South Downs

Climate	+ Rainfall is 950 mm y⁻¹ and is below the UK average + Warm, dry dip slope on the Downs, which face south + Cool, damp scarp slope
Topography	+ Steep scarp slopes + Gently sloping dip slope
Geology soil, drainage and vegetation	+ Underlying chalk provides thin, infertile, well-drained soils which lack minerals such as potassium. These are rendzina soils + The type of vegetation these soils support is slow growing, small and low, for example herbs + Pockets of deeper soils are able to support woodland
Biotic factors	+ There are many different habitats on the Downs, including chalk heath, where acidic windblown deposits overlie the chalk + Acid-loving species such as heathers dominate + Chalk grassland is under threat from changing land use. Many areas are now Sites of Special Scientific Interest (SSSIs)

Ecosystem responses to changes in one or more of their components or environmental controls

Responses to change in the ecosystem of the South Downs include the following.

+ Agriculture:
  + Agriculture has led to the fragmentation of many habitats and the risk of local extinctions.
  + Habitats next to intensively farmed land are affected by spray drift of pesticides and insecticides.
  + Runoff from farmland causes erosion and loss of soil and vegetation.

Now test yourself

20 What is 'litter' in the context of nutrient recycling?

21 Name three nutrient stores.

22 Outline the process of leaching.

23 Why are tropical rainforest soils not very fertile?

Answers on p. 265

TESTED ◯

Revision activity

Draw a mineral nutrient cycle for a tropical rainforest ecosystem and a savanna grassland ecosystem. Add short annotations to explain the connections between climate, soil and vegetation (biomass). For example, in tropical rainforests the soil retains only a small amount of nutrients as fast-growing plants take up the nutrients rapidly.

Exam tip

Scale is an important concept running through this unit. From a puddle to a biome and from a simple food chain to a complex food web, the principles of the structure and functioning of ecosystems can be applied consistently.

129

+ **Leisure activities** such as walking, hang gliding, mountain biking and four-wheel drive rallies disturb rare species and damage the grass cover, leading to erosion of the thin soils.

Factors influencing the changing of ecosystems

Table 6.4 outlines factors that influence the changing of ecosystems.

Table 6.4 The impact of climate change and human exploitation on ecosystems

Climate change	On a global scale, atmospheric warming will lead to many cases of animal extinction. The Intergovernmental Panel on Climate Change says that a 1.6°C rise in temperature may put 20–30% of plant and animal species at risk.
	In the UK, potential impacts include threats to salt marshes as a result of rising sea levels and summer droughts affecting beech woodland. Animal species could also face breeding problems if food is not available at the right time.
	Further possible impacts are shown in Figure 6.11.
Human exploitation	Global population increase (from 3 billion in 1960 to 7.5 billion in 2015), economic growth and advances in science and technology have led to exploitation of the global environment, particularly with regard to: + land-use change for urbanisation and crop growing + deforestation – affecting 8.5% of the world's remaining forest + pollution of oceanic and freshwater ecosystems.

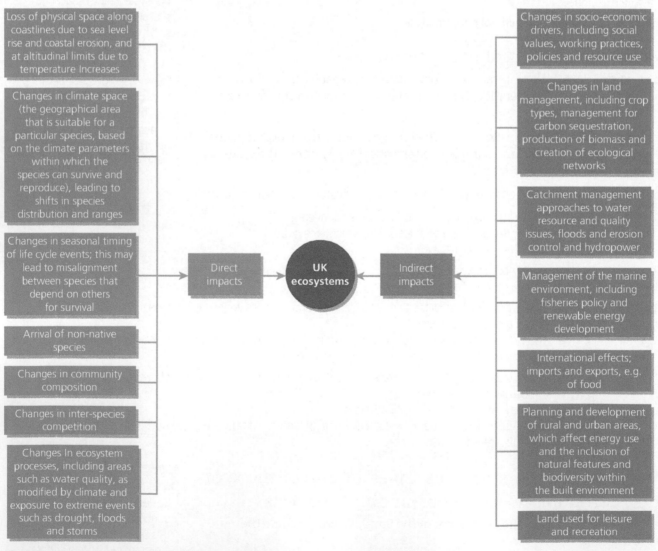

Figure 6.11 Direct and indirect impacts of climate change on UK ecosystems

Making links

See Chapter 7 page 157 for more on the environmental impacts of globalisation.

Check your understanding and progress at **www.hoddereducation.co.uk/myrevisionnotesdownloads**

Revision activity

This is a small-group revision activity.

A range of factors influences ecosystem change. Depending on the size of your group, choose one or two factors from the following list and produce a spider diagram or a flow diagram to explore the direct and indirect impacts of your chosen factor on ecosystems.

+ Increased urbanisation
+ Trade
+ Resource exploitation
+ Increased agriculture (crop growing and animal grazing)
+ Deforestation
+ Leisure and recreation

Figure 6.11 shows an example related to climate change.

Making links

In the above revision activity, links may be made with:
+ trade (Chapter 7 page 152)
+ deforestation (page 131 of this chapter)
+ agricultural change – animal grazing (page 122 of this chapter).

Exam tips

The impacts of climate change refer to a large extent to 'possible futures', which is a key concept for this unit. Make sure that you are able to discuss a range of 'possible future outcomes'.

It is important to practise clear cause-and-effect sequences of explanation to respond to questions on how different factors affect ecosystems. A diagram is useful and will obtain high marks if it is well annotated and detailed. Use subject-specific vocabulary, such as plagioclimax vegetation and seres.

Now test yourself   TESTED

24 Why are changes in seasonal timings a threat to animal populations?

25 How does a prolonged period of drought impact ecosystems?

Answers on p. 265

Biomes   REVISED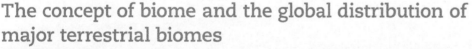

The concept of biome and the global distribution of major terrestrial biomes

A biome is a term used to describe a large-scale ecosystem, usually at a continental scale, which has distinct plant and animal species in the climatic climax stage of succession.

There are eleven terrestrial biomes, which are named after the dominant vegetation. Figure 6.12 shows the distribution of world biomes. They are based upon climate, soils, latitude, topography and native vegetation.

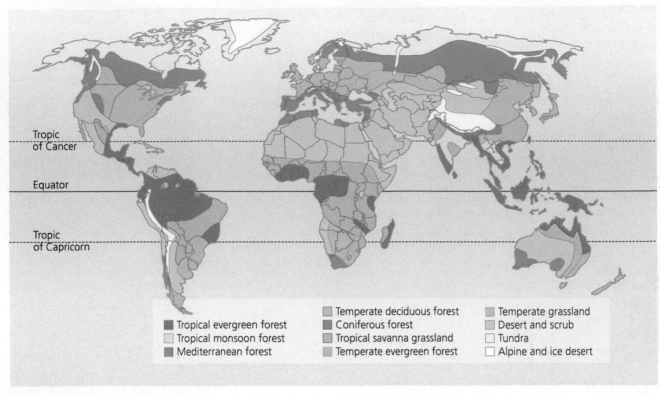

Figure 6.12 World biomes

Now test yourself   TESTED ◯

26 Define the term 'biome'.

27 What is tundra?

Answers on p. 265

Revision activity

Produce a world map to summarise the main terrestrial biomes and annotate with facts relating to climate and plant adaptations.

The nature of two contrasting biomes

Making links

There are links between the human and physical aspects of the two example biomes and:
+ climate change (Chapter 1 page 29)
+ the carbon cycle (Chapter 1 page 21)
+ the soil moisture budget (Chapter 1 page 15)
+ the water cycle (Chapter 1 page 11)
+ trade (Chapter 7 page 148)
+ changing places (Chapter 8 page 169).

Biome 1: tropical rainforest

Figure 6.13 relates to the first biome, a **tropical rainforest** (TRF), and Figure 6.14 shows its layered structure.

Characteristics

Latitude 10°N and 10°S of the equator. Found in Amazon river basin of South America, equatorial part of Africa, SE Asia and Oceania. Climate: average annual temperature of 25°C to 30°C, high annual rainfall of >2000 mm, Convectional rainfall all year round. Soils – latosols, deep, ferrallisation leads to breakdown of bedrock, high concentration of iron and aluminium. Climate gives rise to the highest biodiversity on the planet, with a layered structure (Figure 6.14). Nutrient cycling is fast and nutrients are stored mainly in the biomass, meaning that soils are infertile for agriculture.

Ecological responses

Leaves have a drip tip and waxy surfaces to allow excess rainfall to be shed.

Trees grow rapidly upwards towards the light, resulting in tall, slender trunks and few branches.

Leaves form a dense crown to gain maximum use of sunlight for photosynthesis.

Roots are concentrated close to the surface of soils to make use of rapid nutrient recycling.

Lianas are a common adaptation – thick, woody stems and varying lengths (up to 1000 m), they wind themselves around tree trunks to reach the top of the canopy and extend to other trees.

Epiphytic plants grow on the dark forest floor.

Animal adaptations include camouflage and climbing ability.

Impact of human activity

Mainly forest clearance for:

- shifting cultivation
- commercial exploitation of timber
- cattle ranching
- colonisation programmes
- production of sugar cane, soya beans.

Impacts include:

- disruption of the food chain as habitats and plant species shrink
- leaching of minerals and erosion of topsoil once vegetation is removed
- disturbance of the tropical rainforest microclimate
- local air pollution due to burning
- increased CO_2 levels.

Development issues

Population changes:

- Pressure on indigenous people due to conflict, government policy and human activity.
- Population increase due to variety of human uses and increasing resettlement.

Economic development:

- Mining, forestry, oil exploration and agriculture have all led to development.
- Increased infrastructure.

Agricultural extension:

- Increased food demand has led to two types of agriculture: shifting cultivation and large corporations producing fruit, oil palms, soya and coffee.
- Monoculture.

Biodiversity – the destruction of natural habitats has caused irreversible loss of biodiversity.

Sustainability – reduced-impact logging schemes, debt exchange (some international debt exchanged for forest conservation), ecotourism, deforestation charges.

Figure 6.13 Biome 1: tropical rainforest

Plants	Animals	Metres	Conditions
Epiphytes – plants living in the tree crowns for light (not parasitic)	Emergent layer – birds and insects	50 / 45 / 40	Maximum sunlight, rain and wind: temperatures lower at night
Woody climbers (lianas)			
Dense unbroken cover	Canopy layer – animals living here rarely visit the floor	35 / 30	Trees compete for light: nearly all rain is intercepted
		25 / 20	15% sunlight: rain drips through canopy: hot and humid
Buttress roots for tall trees	Lower layer – animals living in trees here visit the floor	15 / 10	10% sunlight: dark and gloomy, very little change in temperature
Shrubs/herbs		5	
Dense tree trunks, little undergrowth, mosses and ferns	Very few dead leaves on the surface; seeds from trees germinate quickly	0	Warm, moist soil
Shallow root systems			Iron-rich layers (latosol)
			Weathered soil
Leaching of minerals (e.g. calcium)			Weathered rock
			Parent rock
		–100	

Figure 6.14 The layered structure of the tropical rainforest

Biome 2: Savanna grassland

Figure 6.15 relates to the second biome, **savanna grassland**.

Characteristics

Tropical wet and dry climate. Variations in temperature between 18–22°C in wet season and 28–34°C in dry season, depending on location. Rainfall variations of 11 months >1000 mm in wet season to 2 months <500 mm in wet season. Soils laterite/ferruginous, 1–2 m, thin organic matter layer. Active bacteria break down plant matter faster than it is produced. During the dry season, plants die back and litter builds.

Ecological responses

Wetter areas – tall grasses with deciduous trees, which lose their leaves in the dry season, leathery leaves reduce transpiration loss, low crowns shade roots, trees are pyrophytic and can withstand fire; tree savanna, e.g. acacia and baobab.

Grassland savanna – perennial grasses die back in the dry season, grasses are often too sharp or bitter for grazing animals, grasses grow from the bottom up so that growth tissue is not damaged by grazing animals.

Towards desert margins – tussock grass (retain water) and shrub savanna (many acacia trees). Many species have deep roots, short stems; sometimes the stems can photosynthesis so that there are fewer leaves and more water retention.

Animals have long legs or wings to enable long migrations.

Impact of human activity

Mainly grazing activities for cattle, goats, sheep.

Grass is burned off to ensure better growth of young grass next season. The regular burning makes it difficult for young trees and bushes to become established.

Woody plants are killed by cattle eating their foliage. Thorny, animal-repellent trees therefore become more prevalent.

Both above factors lead to a plagioclimax vegetation.

Conservation of the savanna is variable. National parks in South Africa, e.g. Kruger National Park, also Central Kalahari Game Reserve.

Development issues

Nomadic way of life does not support large populations; also people face endemic disease, poor soils and unreliable water supplies.

Expansion into wildlife habitats has resulted in conflicts, e.g. settlers and elephant herds in African savanna. This has led to slaughter of many elephants by shooting, poaching or poisoning.

More humane projects involving the fencing of agricultural areas with bee nests, which frighten the elephants away if they are brushed against.

Population growth outside the savanna is raising food security issues, but growth of crops such as maize and soya is difficult and not carbon efficient.

Some nodes of economic development based on metal ores and diamonds.

Economic development remains in some agriculture and in game conservation, which does attract tourists and creates an income for locals, e.g. in the Serengeti National Park in Tanzania.

The need to grow food puts pressure on the land and reduces biodiversity, while tourism income creates the need for conservation and biodiversity protection.

Figure 6.15 Biome 2: savanna grassland

Revision activity

Create a table to summarise the contrasts between the tropical rainforest and savanna grassland biomes. Remember that contrast refers to similarities and differences.

Exam tip

As with any located example it is essential to be able to support your answers with place-specific facts, for example conservation projects for savanna grasslands – Kruger National Park, South Africa, plant species in grasslands – acacia trees, climatic data – average temperatures of 25–30°C in tropical rainforests.

Ecosystems in the British Isles over time

REVISED

Succession and climatic climax, as illustrated in lithoseres and hydroseres

As stated on page 128, a climatic climax community is the final stage in succession (the process of change in the species structure of an ecosystem over time). It is a state of equilibrium (balance) between plants and animals and their physical environment. However, at any point, succession can be halted by **arresting factors** which can have either a temporary or a permanent impact. The result is a **sub-climax community**. **Plagioclimax communities** are the result of human activity (planned or unplanned), for example burning to clear land for farming, deforestation, natural fires.

Making links

See also page 128 for more on succession.

Check your understanding and progress at **www.hoddereducation.co.uk/myrevisionnotesdownloads**

Table 6.5 and Figure 6.16 show the succession of a **lithosere** and a **hydrosere** in the UK.

Table 6.5 A lithosere in the UK

Bare rock surface	This is initially colonised by bacteria and single-celled photosynthesisers that are able to survive on few nutrients and get most of their energy from the Sun.
	The surface conditions are often dry and the soil little more than particles of weathered rock.
Seral stage 1 Colonisation	The first plant species to colonise an area are called pioneers.
	These are lichens that are adapted to the severe (dry, windy, soil-free) conditions. They begin to break up the rock to form a thin layer of proto-soil.
	As they die they add dead organic matter to weathered rock and windblown dust. This creates a simple soil, which improves water retention.
	Mosses are then able to develop.
Seral stage 2 Establishment	As the soil develops further, ferns and small herbaceous plants and grasses begin to grow.
	Species diversity increases. There are more invertebrates living in the soil, which increase the organic content. This enables the soil to hold more water.
Seral stage 3 Competition	Larger plants, including shrubs and trees, begin to establish themselves. They use up a lot of available water and shade the ground.
	Some of the earlier colonisers are unable to compete and die out.
	Herbivores become established and predators begin to move in.
Seral stage 4 Stabilisation	Fewer new species colonise. Complex food webs develop.
	This stage is dominated by larger, fast-growing trees such as birch and rowan.
	Top predators are found at this stage.
Seral climax	This is the final seral stage. It represents the maximum possible development that a community can reach under the prevailing climatic conditions (temperature, light and rainfall). This is called a climatic climax community.
	The total number of larger plant species falls as a few large species dominate the area. In the case of southern England, these are broad-leaved deciduous trees, such as oak and ash.

Succession in a hydrosere has five stages: 1) open water, 2) rooted plants, 3) swamp stage, 4) marsh stage, 5) carr and woodland stage. They are explained in Figure 6.16.

Stage 1: Deep freshwater will not support rooted submerged plants as there is insufficient light. There will be floating blue and green algae.

Stage 4: Swamp plants that grow in partly submerged conditions gradually die out as the pond edge rises above water level. Yellow iris continues, water mint adds scents and seedlings begin to grow.

Alder/willow carr. The ground is no longer completely saturated and anaerobic. Willows and alder trees that like moist ground dominate and they shade out the marsh undergrowth. Woodland floor plants like ferns and sedges appear.

Stage 2: Sediments are deposited in the pond reducing its depth. There will be rooted submerged plants like pondweed or rooted plants like lilies with floating leaves.

Stage 3: By this swamp stage the water may be shallow enough to allow emergent plants like yellow iris, branched bur-reed and greater reedmace.

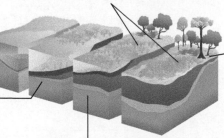

Figure 6.16 A hydrosere in the UK

The characteristics of the climatic climax: temperate deciduous woodland biome

A temperate deciduous biome is found in north-west Europe, for example the British Isles, and also the USA and South Island, New Zealand.

Oak can reach heights of 30–40 m and is the dominant tree species as the climax vegetation develops. The maximum number of tree species per km² in the British Isles is eight. Some woodlands may have a single dominant species, for example beech.

Woodlands show stratification as they develop to a climatic climax (see Figure 6.17).

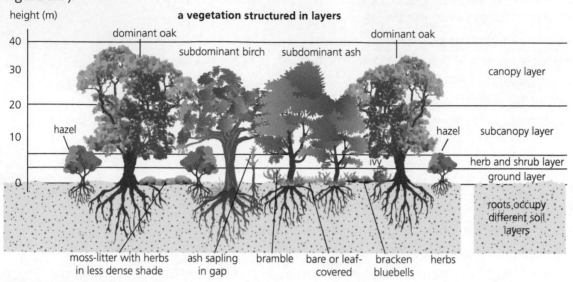

Figure 6.17 Woodland stratification

The effects of human activity on succession

Figure 6.18 outlines the effect of human activity on heather moorland.

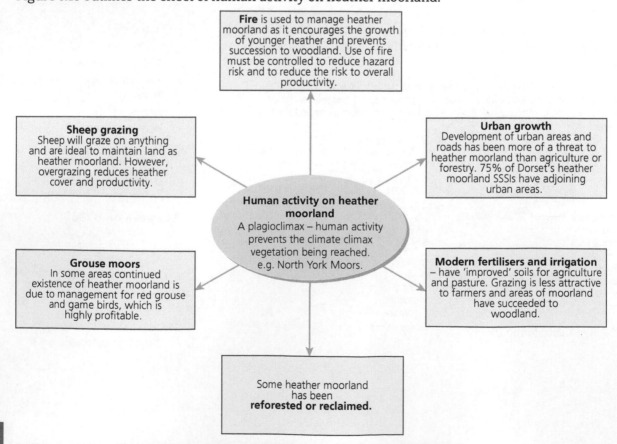

Figure 6.18 The effect of human activity on heather moorland in the British Isles

Check your understanding and progress at **www.hoddereducation.co.uk/myrevisionnotesdownloads**

Making links

See Chapter 1 page 19 for more on the effects of localised, improved farming practices and page 24 for the effects of fire on the carbon system.

Marine ecosystems

The distribution and characteristics of coral reef ecosystems

+ Tropical coral reefs are found between 30° north and 30° south of the equator where surface water temperatures do not drop below 16°C.
+ Such reefs can form structures as long as the 2000 km Great Barrier Reef.
+ Coral reefs are composed of layers of calcium carbonate or limestone. The living reef lies over the top.
+ The reef is built by coral polyps and coralline algae.
+ The algae need sunlight for photosynthesis and so tropical corals grow only where the water is shallow and clear.
+ The algae give the corals their colour.
+ If the water temperature becomes too high, the algae leave the polyp. This exposes the white calcium carbonate skeletons of the coral, a process which is referred to as **coral bleaching**.

There are three main types of coral reef:
+ **fringing reefs** – grow directly from the shore
+ **barrier reefs** – extensive reefs parallel to the shore
+ **atolls** – rough, circular reefs surrounding a lagoon.

> **Exam tip**
>
> When discussing the impacts of human activity on coral reefs, be clear on the location, type of reef and nature of the impact – do not write in generalised terms.

> **Now test yourself**
>
> 31 Where are coral reefs typically found?
> 32 What is coral bleaching?
>
> **Answers on p. 265**
>
> TESTED

An example of a named and located coral reef ecosystem – the Jamaican coral reef

The Jamaican coastline has a coral reef of 1240 km², mostly fringing reef along the north and east coasts (Table 6.6).

Table 6.6 Jamaica's coral reefs

Natural features	Human activity and its impact	Future prospects
+ Coral bleaching of 34% of the coral, caused by unusually warm sea surface temperatures (1°C above norm is enough; increase has been as much as 2°C). + Best coral where the salinity of the seawater is 34–37 parts per 1000. In the mouth of the South Negril River and Rio Cobre in Jamaica there is a fall in salinity and a gap in the coral. + In the 1990s, a disease epidemic led to mass mortality of sea urchins, which in turn led to rapid algal growth. This limited sunlight and caused a reduction in O_2. Marine life and coral reefs were seriously affected. + Major hurricanes have damaged Jamaica's coral reefs. Strong waves fragment and kill branching coral in shallow water.	+ Pollution – from onshore bauxite mining, the sediment reduces water clarity and much-needed sunlight. + Tourist developments have led to sewage being discharged into the sea with little treatment. + Agricultural fertilisers discharged into the marine environment add nutrients, which lead to eutrophication and algal blooms. + Over-fishing has resulted in an 80% reduction in fish biomass. + There are concerns regarding the potential impacts of a desalination plant, which would release chemicals, causing localised increases in salinity and temperature.	+ In the Caribbean the proportion of reef covered by coral fell from 50% in the 1970s to 8% in 2013. + The loss of coral reefs impacts the tourist industry. + Corals buffer the impacts of incoming storms and to lose them would lead to more storm damage. + Climate change leading to coral bleaching is a major future threat to one of the world's most diverse ecosystems. + A solution to the threats may be to set up underwater national parks. + Marine-protected areas are increasing on the coast of Jamaica and now account for 15% of the country's coastal water. This will no doubt increase in the future.

Now test yourself   TESTED

33 Explain the process of eutrophication.

Answer on p. 265

Exam tip

When evaluating future prospects, remember to include positive measures, such as conservation strategies, as well as negative impacts.

Local ecosystems   REVISED

A distinctive local ecosystem – a pond

Exam tip

Do not just learn about impacts on an ecosystem. You must also address responses to the impact by different stakeholders.

Characteristics
+ Small, shallow, permanent or seasonal water bodies up to two hectares in size.
+ Abiotic factors include temperature, water flow, salinity and dissolved oxygen.
+ Biotic factors include reeds, bulrushes, frogs, toads and newts.

Ecological responses to climate and adaptations of flora and fauna

A pond ecosystem consists of four habitats:
+ **Shore:** sandy shorelines allow the growth of grasses and algae, and allow earthworms, snails and insects to thrive.
+ **Surface film** is inhabited by free-floating organisms.
+ **Open water** supports carp, pike, plankton and zooplankton.
+ **Pond bottom** varies according to the depth of the water. Shallow water provides a breeding environment for snails and insects; deeper ponds have flatworms and dragonfly nymphs.

Exam tip

When discussing the human impact on ecosystems it is worth considering the source of the impact, that is, is the impact from indigenous societies or as a result of advanced technology?

Now test yourself   TESTED

34 Why is it important to maintain ponds as a feature of the countryside?

Answer on p. 265

Figure 6.19 shows the food web of a pond.

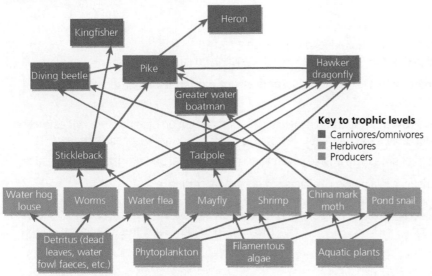

Figure 6.19 Generalised food web for a pond

Local factors in ecological development and change

The development of a hydrosere, such as a pond, is shown in Figure 6.16 on page 136.

The impacts of change and measures to manage these impacts

Due to a combination of change in agricultural practices and land-use change, ponds have gradually disappeared from the countryside of the British Isles.

The **Million Ponds Project** aims to create an extensive network of ponds once again. Measures to achieve this include:
+ funding for new ponds
+ technical support for pond creation
+ raising the profile of ponds through media and policymakers.

Phase II of the project involves:
+ identification of important areas for ponds (IAPs)
+ targeting landscape-scale pond creation
+ applied research projects.

Case studies

A specified region experiencing ecological change, for example the Sundarbans, Bay of Bengal

This case study should:
+ illustrate and analyse the nature of change and the reasons for it, how the economic, social, and political character of its community reflects its ecological setting
+ illustrate and analyse how the community is responding to change.

What do I need to know?	Content and suggested revision methods

The nature of the ecological change
 	
The reasons for it
	Revision notes on:
	+ the nature of change:
	  + decline of native plant species, forest being submerged
	  + depletion of the biodiversity in the area
	  + retreating coastlines
	  + rising sea temperatures impacting on the fragile ecosystem
	+ reasons for it:
	  + global climate change leading to rising sea levels
	  + over exploitation of aquatic species

	+ increasing occurrence of tropical storms + illicit logging + deforestation + intensification of agriculture due to population growth + diversion of water from the Ganges leading to increased salinity. Organise revision notes into a series of **cause-and-effect explanations** based on the lists above, e.g.: **Rapid population growth** in India and Bangladesh → intensification of agriculture → destruction of mangroves in Sundarbans → coast left vulnerable to tropical storms → coastline retreats.
How the economic, social and political character of the community reflects the ecological setting	Organise this section of notes into a table covering economic, social and political characteristics. For each characteristic, explain how it reflects the ecological setting, e.g.: | Characteristics	How it reflects the ecological setting | |---|---|
Social and economic Subsistence agriculture – most families are dependent on subsistence agriculture	This has led to increased pressure on a fragile ecosystem.	
How the community is responding to change	Responses include: + designation of the ecoregion as a World Heritage Site + restricting fishing. List the responses. Select 3–4 responses and categorise them as national government, international agencies, community responses. Make summary notes on the different responses.

A specified ecosystem at a local scale, for example the Sefton coast sand dunes

This case study should:
+ illustrate and analyse the key themes set out in this unit, including the nature and properties of the ecosystem and the human impact upon it
+ illustrate and analyse the challenges and opportunities presented in its sustainable development.

What do I need to know?	Content and suggested revision methods
The nature and properties of the ecosystem	The best way to revise this section is to produce a **cross section** of the **dunes** and add detailed annotations which include: + a description of each section of the dunes – embryo dunes, mobile dunes, semi-fixed and fixed dunes, dune slacks and climax vegetation + information on plant succession. You might need to use A3 paper. Annotations can include fieldwork data.
The human impact on the ecosystem	Make summary notes on the human impacts, e.g.: + dumping of 'night soil' and its impact of flattening the dunes + uncontrolled public access + golf course management.
Challenges and opportunities for sustainable development	Sustainable development of the area will require a careful balance between preserving and protecting the natural ecosystem and allowing public access and usage. In a table, summarise the management and conservation initiatives and highlight which aspects of these initiatives create an **opportunity** and which present a **challenge**. For example: Projects by the National Trust to recolonise the dunes with native species are challenged by the loss of dunes due to increased marine erosion. Other management/conservation initiatives to cover in the table include: + planting of pine trees + setting up of the Ainsdale National Nature Reserve + conservation of fixed dunes.

Check your understanding and progress at **www.hoddereducation.co.uk/myrevisionnotesdownloads**

Exam practice

1 Explain the plant adaptations of savanna grasslands. [6]
2 Compare and contrast mineral nutrient cycles in a tropical rainforest and a savanna grassland ecosystem. [6]
3 Assess the impact of human activity on succession for a named ecosystem in the British Isles. [9]
4 To what extent are humans responsible for declining global biodiversity? [9]
5 With reference to a region experiencing ecological change you have studied, to what extent are the changes the result of human activity. [9]

Answers and quick quizzes online

Exam skills

Opportunities to practise geographical skills within this topic include:
+ observation skills in fieldwork
+ collection, manipulation, presentation and analysis of primary and secondary fieldwork data – including quantitative and qualitative sources
+ statistical tests applied to data (for example primary and secondary sources in fieldwork)
+ analysis of a range of graphs and histograms showing, for example, climatic data, species diversity and change
+ interpretation of geospatial data (for example maps on the distribution of global biomes and ecosystem degradation)
+ simple data manipulation using basic statistical manipulation tests such as mean, median, mode, frequency distribution, measures of dispersion and more advanced statistical tests such as student t-test to compare species concentrations in different locations.

Summary

+ Understand the concept of biodiversity and the local and global trends in loss of biodiversity.
+ Be clear on the link between ecosystems and human populations – for example, in what ways do humans depend on ecosystems?
+ It is important to learn the subject-specific vocabulary in order to accurately explain the structure and functioning of ecosystems.
+ Ecosystems function through a sequence of energy flows, food chains, food webs and nutrient recycling.
+ The structure and functioning of ecosystems must be applied to a range of specific examples at different scales.
+ Ecological succession occurs gradually, in stages, and eventually may result in stable, climatic climax vegetation. However, there is a range of arresting factors that often prevents this.
+ Understand the concept of a biome and be able to compare and contrast the features and functioning of two examples: the tropical rainforest and the savanna grassland.
+ Development of ecosystems specific to the British Isles must be covered, for example a lithosere, a hydrosere, temperate deciduous woodland and a plagioclimax, such as heather moorland.
+ Coral reefs form a fragile ecosystem where human activity is posing a significant threat to future prospects.
+ In addition to learning about the nature of human impacts on ecosystems and the responses to them, the prospects for a sustainable future should be considered.

7 Global systems and global governance

Globalisation

Dimensions of globalisation

Globalisation:
+ is a process that integrates people across the world
+ includes any process of change that is global in its occurrence and impact
+ has a range of negative and positive impacts.

Links to globalisation can be **environmental**, **economic**, **social**, **cultural**, **political** and **technological**. The focus of globalisation has been primarily on its economic dimensions; however, it encompasses a wide range of further dimensions, as illustrated in Figure 7.1.

> **Globalisation** The process by which the world is becoming increasingly interconnected.

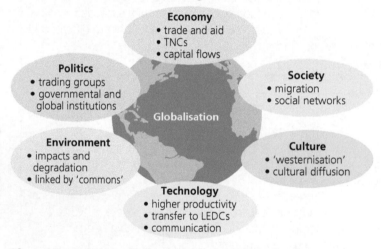

Figure 7.1 Factors and dimensions of the globalisation process

> **Exam tip**
>
> Globalisation is a complex process that drives not only economic but also social, cultural, environmental and political change.

Flows of capital

Capital can involve the physical transfer of production resources but usually refers to the flow of money for investment, trade or production.

Deregulation led to a removal of restrictions on the movement of capital.

Models on capital flows (for example Frank's 'dependency theory') assume that:
+ global power is concentrated in a 'core' of highly developed economies (**HDEs**) and
+ 'peripheral', less developed economies (**LDEs**) suffer a lack of investment, leakages and outmigration.

Emerging economies in BRIC (Brazil, Russia, India and China) and MINT (Mexico, Indonesia, Nigeria and Turkey) groups illustrate a continuum of development.
+ **Foreign direct investment** (FDI) is an investment made by a government or large company into the physical capital or assets of a foreign enterprise. The US, the EU and China are the main recipients of FDI, while Japan is the country with the highest outflows.
+ **Repatriation of profits** is the process whereby profits made by a company abroad are returned to the home country. This is often a flow back to a developed country, termed a leakage from the economy of the host country.
+ **Aid** refers to financial support for poorer countries, mainly from rich countries. It can have many forms:
  + multinational aid provided by many countries

> **Capital** Accumulated wealth, which may include machinery, buildings, money, stocks or investments.
>
> **Deregulation** The removal of government rules, regulations and laws from the operation of business.
>
> **Leakage** The economic loss of profits back to companies owned outside of the host country.

Check your understanding and progress at **www.hoddereducation.co.uk/myrevisionnotesdownloads**

- a bilateral agreement between two countries
- emergency aid provided by an NGO or the Red Cross
- development aid.

+ It may have conditions (jobs, political support).

+ It has advantages and disadvantages to the receiving nation (it provides humanitarian relief, but can promote dependency rather than self-reliance).

+ **Remittance payments** refer to the money workers in foreign countries transfer back home to their family. This is now the second most important form of income to developing nations (see Figure 7.2). The incoming cash boosts the economy of the recipient country and provides hard currency.

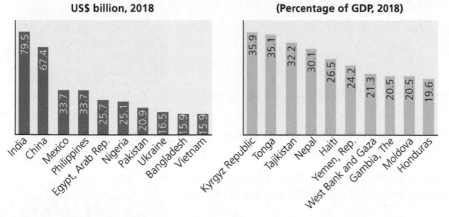

Figure 7.2 The world's top 10 remittance receivers 2018, by total amount and as a percentage of GDP

Figure 7.3 summarises the main flows of finance and capital between regions in the global economy.

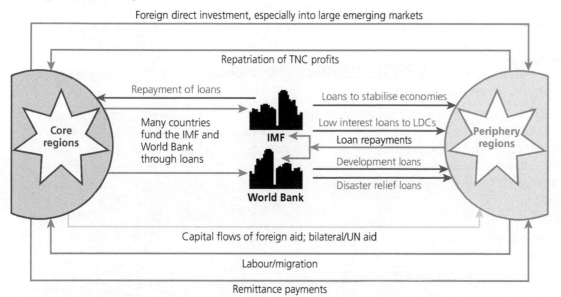

Figure 7.3 Flows of finance and capital between regions in the global economy

Making links

See Chapter 8 for more on the impact of globalisation on place identity and representation, particularly in terms of culture.

Flows of labour

Labour is the human resource available in the economy. It is less mobile than capital but there have been significant increases in global economic migration.

Although the pattern is constantly changing, in broad terms, the main flows are:
+ from Latin America to North America
+ between neighbouring countries in Africa
+ between European nations
+ from Africa and south Asia to Europe.

143

Most movements are within geographical regions and between neighbouring countries. Most migrants travelling longer distances are those with education and financial means.

Flows of products

Products are tangible articles manufactured for sale.

Increased flows of goods have been facilitated by a reduction in transaction costs as a result of the improved flow of data and payments.

Containerisation has improved the flow of goods over long distances, while commercial jet aircraft and super freighters have also had an impact.

The oversight of the WTO has helped the flow of goods through the reduction of protectionist measures such as tariffs.

Flows of services

Services refer to the provision of a service rather than goods, for example banking, transport and retail.

Footloose services such as banking and advertising can locate anywhere thanks to improvements in flows of information and capital.

There has been a decentralisation of some **low-level services** (banking, travel, tourism) to developing countries, for example call-centre operations and tourism.

High-level services (for example financial and investment services) are concentrated in cities in the developed world, such as London, New York and Tokyo.

A growing number of conglomerates have emerged seeking to extend their influence at a global scale (for example HSBC).

Flows of information

Information is wide ranging and can involve the transfer of cultural ideas, language, technology and design, among much else.

Digitisation and satellite technology have transformed the flow of information, and mobile telecommunication technology is more widespread:
+ Email and internet use enables the fast flow of large amounts of information and enhances business communication.
+ Satellite technology allows live media coverage across the globe.

Now test yourself
TESTED

6 Define the following: a) leakage, b) remittance, c) footloose, d) containerisation, e) deindustrialisation.
7 Differentiate between high-level and low-level services.

Answers on p. 265

Global marketing

For global companies marketing is now seen as a process operating across one single market. They establish a recognisable brand and adapt it across global regions. Many fast-food giants have followed this principle, with menus adapted for tastes in growing markets in India and China, for example.

Patterns of production, distribution and consumption

Production

Production has decentralised over time to developing economies, mainly due to lower costs of labour and land. The production side of many transnational corporations (TNCs) now takes place away from the home nation where the head office is located.

Containerisation A system of standardisation that uses large, standard-sized containers for transport.

Protectionist measures Policies of erecting barriers to trade, for example quotas and tariffs.

Tariffs Taxes on imported goods.

Conglomerate A business made up of a number of other, sometimes unrelated, businesses.

Footloose An industry that can be placed in any location and is not affected by factors such as resources and transport.

Revision activity

Create a revision spider diagram that summarises the main features of flows of capital, labour, products, services and information in the global economy. You can add information from your own notes. Remember to give examples wherever possible.

Exam tip

It is important to be able to refer to located examples and major flows, for example remittance income in Figure 7.3.

Check your understanding and progress at **www.hoddereducation.co.uk/myrevisionnotesdownloads**

Factors affecting locational decisions of large manufacturing companies include:

- availability of a skilled workforce
- opportunities for high-tech production plants
- tariff-free markets
- infrastructure support.

One consequence of this global shift has been deindustrialisation and a loss of manufacturing jobs in richer countries.

Deindustrialisation The reduction of industrial activity in a region or country.

HDEs have reacted to deindustrialisation with strategies such as:

- offering incentives, for example as tax breaks, for TNC investment
- upgrading skills and technology
- protectionist policies, for example import tariffs.

Distribution and consumption

Distribution of manufactured goods is now more versatile and far reaching due to the advances in transport and communication networks referred to on page 145. Transport time and relative costs have fallen dramatically.

Consumption is mainly still in the developed economies. However, a major trend is the growing middle class of industrialising countries, such as India, Brazil and China, which are more affluent and have growing spending power, demanding more consumer products.

Now test yourself

8 How are some HDEs reacting to deindustrialisation?
9 Why is consumption increasing in countries such as India and Brazil?

Answers on p. 266

TESTED

Factors in globalisation

A number of factors have been responsible for the drive towards globalisation – these are summarised in Figure 7.4.

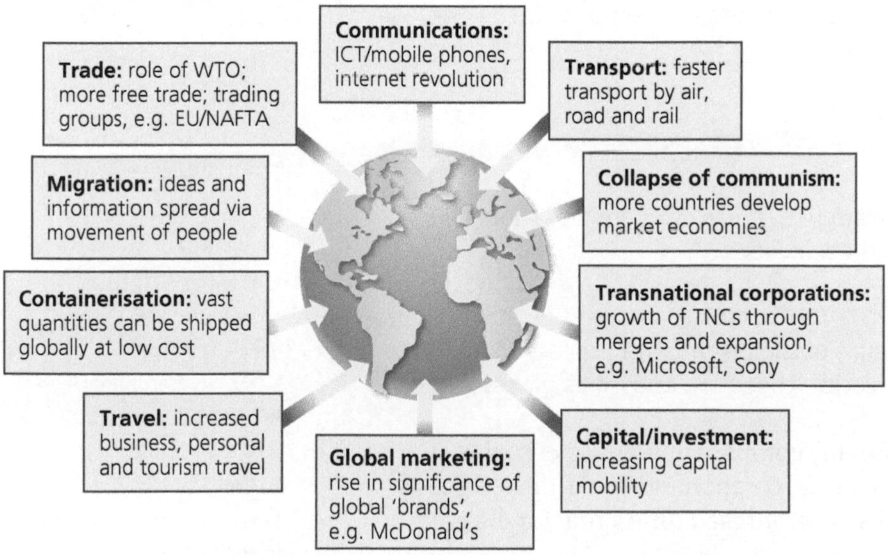

Trade: role of WTO; more free trade; trading groups, e.g. EU/NAFTA
Communications: ICT/mobile phones, internet revolution
Transport: faster transport by air, road and rail
Migration: ideas and information spread via movement of people
Collapse of communism: more countries develop market economies
Containerisation: vast quantities can be shipped globally at low cost
Transnational corporations: growth of TNCs through mergers and expansion, e.g. Microsoft, Sony
Travel: increased business, personal and tourism travel
Global marketing: rise in significance of global 'brands', e.g. McDonald's
Capital/investment: increasing capital mobility

Figure 7.4 Factors that have accelerated the process of globalisation since the 1990s

Exam tip

Develop an awareness of the relative importance of factors that have accelerated the process of globalisation. For example, do ICT communications have more impact than personal business travel?

Certain technologies, systems and relationships have also influenced the process of globalisation.

Financial systems

- International trading is now faster and easier than ever before.
- The **global financial system** (GFS) provides a framework to facilitate flows of capital.

+ Electronic trading systems mean that companies around the world can trade rapidly and securely.
+ **Deregulation** of financial markets has led to the removal of barriers to the movement of finance.

Transport systems

+ The increased size and standardisation of many containers used to transport goods has meant that transport of manufactured goods is now more efficient.
+ Reductions in cost, computerised logistics systems and data analysis of efficiency of handling and distribution have also facilitated freer movement, for example the use of standardised containers for sea, rail, road and air transport.

Security systems

Many security issues challenge an increasingly globalised economy. The main ones include:
+ organised crime
+ terrorism
+ bio-security
+ fiscal security
+ smuggling (consider, for example, the work of the World Customs Organization)
+ cybercrime.

Communications technologies

Advances in technology are one of the main reasons for rapid globalisation.

Communication links between countries have grown because of:
+ development of computer technology
+ expansion of the internet
+ increased usage of mobile phones
+ robotic technologies
+ computerised logistics systems
+ innovations, which have become smaller, more efficient and more affordable to both individuals and businesses.

Management and information systems

+ It is now common for various stages of the production process to be located in different parts of the world – these are known as **global value chains**.
+ The global production network for any company involves a complex system of flows of information, finance, components and finished goods. These global production networks are organised differently for different industries and as costs vary.
+ Just in time (JIT) technology has also enabled a range of cost-saving advantages to the production of goods, for example part-assembled goods ready to be quickly finished and distributed as and when required. Industries using JIT technology include car manufacture and the computer industry.

Now test yourself

10 How have financial systems contributed to the process of globalisation?

11 How have computerised logistics systems aided the flow of goods?

Answers on p. 266

TESTED

Trade agreements

These can be **bilateral** (between two countries) or **multilateral** (between three or more countries).

Trade groupings can take the form of:
+ free trade areas
+ customs unions
+ common markets
+ monetary unions.

Figure 7.5 summarises some of the main global trading blocs.

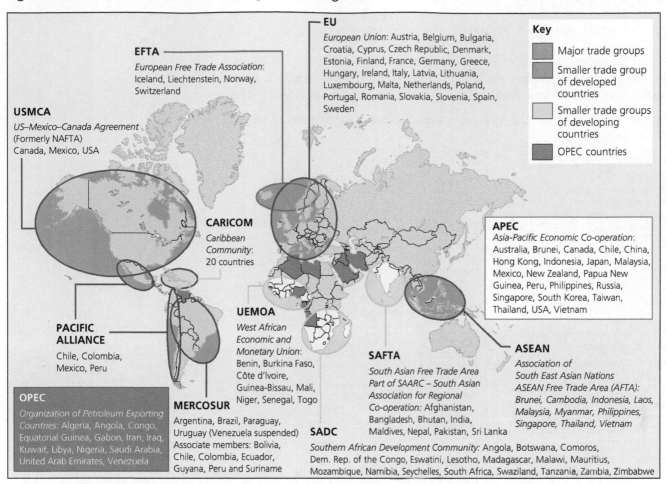

Figure 7.5 Some of the main global trading blocs

Table 7.1 summarises the advantages and disadvantages of trade agreements.

Table 7.1 Advantages and disadvantages of trade agreements

Advantages	Disadvantages
Economic development, through the **economic multiplier**	Lack of access to trading blocs by poorer nations, widening the development gap
Intergovernmental support and security	Trade disputes arising over tariffs, prices of commodities and changes in trade agreements
Representation and influence in world affairs	Border and customs authorities facing corruption and breaches of security
Freedom of movement of goods and labour	Some loss of sovereignty
Easier negotiation of trade with other large trading groups	Pressure to adopt central legislation
Sharing of technological advances	

Now test yourself

TESTED ⬤

12 What is a global value chain?

13 State two advantages and two disadvantages of trade agreements.

Answers on p. 266

Exam tip

You need to know the major trade groups and be aware of smaller trade groups in developed and developing countries. It is important to be able to evaluate the pros and cons of trade agreements.

Economic multiplier
The process by which growing economic activity in an area creates employment; employees have money to spend on goods and services and more economic growth is stimulated.

My Revision Notes: AQA A-level Geography Second Edition

Global systems

Form and nature of interdependence in the contemporary world

A number of systems have evolved to support the economic, political, social and environmental interdependence that has resulted from globalisation.

Figure 7.6 summarises the four main types of interdependence in the modern world.

Economic interdependence

Trade
- More nations participating
- Countries rely on other countries to supply the goods and services they need
- Trading relationships are vulnerable to a nation's political actions

Advances in technology
Connections between foreign businesses have brought new innovations

Employment
There are changes in where jobs are generated and lost

International migration
There is a more mobile global workforce

TNCs and investment
There is growth in investment by TNCs in developing countries

Supply chains
Global supply chains have been formed for individual products

Industrialisation
There has been industrialisation of some countries (e.g. Brazil and India)

There has been deindustrialisation of others (e.g. parts of Europe and North America)

Political interdependence

Intergovernmental organisations give support to global systems (e.g. IMF, WTO, UN)

Social interdependence

Health
Organisations such as the WHO give support in times of global pandemics (e.g. COVID-19)

Education
Existence of student foreign exchange programmes (e.g. Erasmus in Europe)

Culture
Migration strengthens social ties between countries

Environmental interdependence

Global commons
Interdependence generated by shared use of resources

Global climate change
There is now a more international approach led by agencies such as the UN Framework Convention on Climate Change

Unsustainable practices
These challenge environmental interdependency, e.g. air pollution and acid rain

Figure 7.6 The four main interdependencies in the modern world

Revision activity

Either from your class notes or from additional research, make bulleted revision notes on a) the role and b) the criticisms of the following global organisations:
+ the International Monetary Fund (IMF)
+ the World Trade Organization (WTO)
+ the United Nations (UN).

For example:

Organisation	Role	Criticisms
IMF	To stabilise international exchange rates and facilitate development	+ A lack of concern for democracy + Slow to react to crisis situations + Mostly controlled by HDEs
WTO		
UN		

Now test yourself

14 How is economic interdependence reflected in trade?

15 Why is air pollution classified as a form of environmental interdependence?

Answers on p. 266

TESTED

Exam tip

Evaluation of global institutions is important. Remember that they do have positives: a coordinated response, collective decision making, independent advice.

Issues associated with interdependence

Unequal flows

Unequal flows of **people, money, ideas** and **technology** within global systems can sometimes act to promote stability, growth and development, but can also cause inequalities, conflicts and injustices for people and places.

Check your understanding and progress at **www.hoddereducation.co.uk/myrevisionnotesdownloads**

Revision activity

Make and complete a copy of Figure 7.7, which summarises the positive and negative effects of unequal flows. The first part on 'people' has been completed for you as an example. Add boxes and your own notes for money, ideas and technology.

People		Money	
Positive effects	**Negative effects**	**Positive effects**	**Negative effects**
• When there is a lack of employment in one area, work can be sought elsewhere. • Labour migration addresses skills and labour shortages in some areas. • Remittance payments sent back to a home country can boost its economy. • When migrant workers pay taxes and spend money it boosts the economy of the country they are staying in. • Some workers return to their country of origin with new skills and ideas.	• Countries may lose their best talent as labour is attracted away. • Loss of skilled workers impacts on productivity and growth. • Developing countries become over dependent on remittances. • Families are separated as workers depart. • Resentment can build towards migrants in some countries. • Greater movement of the workforce produces a risk of the spread of disease.		

Effects of unequal flows

Figure 7.7 Positive and negative effects of unequal flows

Now test yourself

TESTED ◯

16 In what way could remittances harm a country?

17 How can migrant workers benefit a host country's economy?

Answers on p. 266

Inequality issues

As a result of globalisation, inequalities can exist both within and between countries.

Inequality between countries

✚ Globalisation is reducing global inequality through the transfer of capital and income from rich to poor countries.
✚ The development continuum is more condensed.
✚ Some very poor countries still lag behind – consider Chad and Burkina Faso.
✚ The fastest growing economies are in Asia and Sub-Saharan Africa.
✚ Africa is showing advances in terms of growth but still contains the majority of the very poorest countries.

Inequality within countries

✚ Within many advanced countries inequality has worsened, for example Britain and Canada.
✚ In China and Sub-Saharan Africa inequality is reducing – the gap between rich and poor is narrowing.
✚ The majority of people live in a country showing an increasing poverty gap. The main exceptions are Latin American countries, such as Mexico.

Exam tip

With regards to inequality **within** countries, make sure you have located examples to refer to and that you fully understand the nature of the inequalities. Students often overlook this area but it could be the focus of an exam question.

149

+ The richer in society can potentially cope better with the changes needed in skills and technology and therefore do not become disadvantaged in the job market.

Making links

Inequalities within countries can prompt rebranding and reimaging strategies as a response to negative place representation and perception. See Chapter 8 page 170.

Now test yourself　　　　　　　　　　　　　TESTED ◯

18 Why are the richer in society more able to prosper further?

19 Suggest three reasons why some poor countries continue to lag behind.

Answers on p. 266

Unequal power relations

+ There is a view that the economic integration brought about by globalisation has decreased the likelihood of conflict between nations. However, there is still a feeling of marginalisation of poor countries as it is the rich countries that drive globalisation.
+ The growth in global communications can spread conflict. Consider the Arab Spring of 2010 spreading to North Africa and the Middle East.
+ Countries often use trade to enhance their position of power in a conflict situation, for example sanctions brought by the USA and Europe against Russia for the conflict with Ukraine.
+ HDEs provide aid, investment and transfers of technology to developing countries but often require geopolitical support in return.
+ Wealthier countries have close relations with each other in groups such as the G7.
+ Developed countries have more influence in global governance through groups such as the IMF.

Revision activity

Sometimes nation states use their political and/or economic power in the global system to influence geopolitical events to their own advantage.

Based on an example you have studied, make revision notes in the form of a diagram, table or annotated base map on the:
+ nature of the influence
+ advantages to be gained by the country exerting influence
+ negative consequences of the action/influence.

Now test yourself　　　　　　　　　　　　　TESTED ◯

20 Why is an overdependence on foreign investment not good for developing countries?

21 State one advantage and one disadvantage of groups such as the G7.

Answers on p. 266

International trade and access to markets　　REVISED ◯

Revision activity

It is important to understand some of the key terms relating to the topic of international trade.

Complete Table 7.2 with definitions either from your class notes or from further research.

Table 7.2 Key terms relating to international trade

Term	Definition
Comparative advantage	
Multilateral trade agreement	
Tariff	
Import licence	
Import quota	
Subsidy	
Sanction	
Embargo	

Global features and trends

General features

Figure 7.8 shows the pattern of international trade across the world. A geographical feature of the volume and pattern of international trade is its unevenness.

Trade is in the main dominated by the more advanced and the rapidly emerging economies, which have the economic wealth and political control to advance their trade position. The least developed countries have limited access to global markets and much weaker terms of trade.

Exam tip

Make sure you can identify and name the flows that play a key role in globalisation.

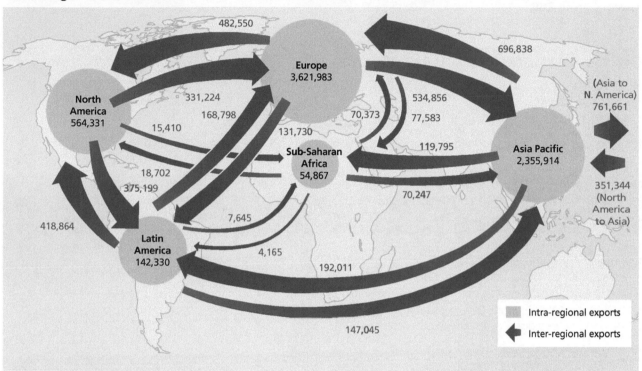

Figure 7.8 Patterns of intra-regional and inter-regional world trade, 2016 (US$ millions)

Factors driving current patterns, trends in trade and trends in investment

The factors driving current patterns are:

+ **comparative advantage** – countries focus on exporting goods that they can produce at a lower cost
+ **proximity** – countries are more likely to trade with their neighbours
+ **agglomeration** – clustering of industries in a certain area
+ **market size and strength** – exporters are drawn to large, affluent markets
+ **geopolitical relationships** – trade follows political alliances.

Trends in trade

+ Trade is dominated by large economic blocs (e.g. North America, Europe, Asia-Pacific region).
+ China is the world's largest trader.
+ Intra-regional trade is strong within Europe and the Asia-Pacific region.
+ Less intra-regional trade takes place in Latin America and Sub-Saharan Africa.
+ Latin America has the strongest trade flows with North America.
+ North America (mainly the USA) has a trade deficit with other regions.
+ Asia-Pacific has a trade surplus with other regions.
+ Intra-regional trade is increasing between the Asia-Pacific region, Sub-Saharan Africa and Latin America.

Trends in investment (foreign direct investment)

The trends in foreign direct investment (FDI) include:

+ Japan, China and France are the largest FDI investors.

Making links

See page 143 for a definition of FDI and more on capital flows.

See page 149 Global systems and page 157 Global governance for further information on world trade organisations.

+ 40 per cent of FDI flows come from the Asia-Pacific region.
+ There has been a slump in FDI from North America.
+ Of the top 20 recipients of FDI, eight were developing economies.
+ Developing countries receive almost twice as much FDI as they initiate.

Now test yourself

TESTED ◯

22 Define comparative advantage.

23 Why do some countries, such as the USA, have a trade deficit with other countries?

24 Why do developing countries receive much more FDI than they initiate?

Answers on p. 266

Trading relationships and patterns

Revision activity

Table 7.3 sets out a summary of current trading relationships and patterns.

Information on HDEs has been added. Complete the table by writing your own bulleted revision points for emerging economies and LDEs.

Table 7.3 Trading relationships and patterns

Highly developed economies	Emerging economies	Less developed economies

USA
+ Traditionally a protectionist country.
+ Up to 2020, the Trump administration favoured protectionist policies, e.g.:
  + USA withdrew from the Trans-Pacific Partnership negotiations
  + the Transatlantic Trade and Investment Partnership collapsed in 2016
  + USA pursued bilateral trade with other countries.
 	
EU
+ The EU negotiates trade deals as a bloc.
+ 65% of trade in the EU is intra-regional.
+ Bilateral trade deals do exist (e.g. with Australia, New Zealand and Japan).
+ Recent trade deals include: Canada–EU 2017, EU–Mercosur (South American trading bloc) 2019 (awaiting full ratification), post-Brexit EU–UK trade deal.
	China
	
	India
		Latin America
		
		Mercosur
		
		Pacific Alliance
		
		Sub-Saharan Africa

Now test yourself

TESTED ◯

25 How does international trade promote economic growth?

26 In what ways can trade a) cause conflicts, b) be used as a bargaining tool within a conflict situation?

Answers on p. 266

Differential access to markets

Special and differential treatment (SDT) agreements are a defining feature of the global trading system. The United Nations Conference on Trade and Development (UNCTAD) aims to keep issues supporting the position of poorer economies of the world, which may be outside the major trading blocs, at the forefront of WTO talks.

Special support measures for least developed countries have been put in place to:
+ address the concentration on the export of primary goods, which are vulnerable to price volatility on global markets
+ provide incentives for export diversification
+ promote faster income growth and achieve economic take-off.

Exam tips

Make sure that your facts regarding trading relationships are up to date. For example, the political events of 2016 with the Brexit vote and future trade deals with the EU and beyond, and the election of Joe Biden as US president in 2020 will redefine key trading relationships.

The key features and patterns of global trade are constantly changing. Make sure that your facts are up to date. Useful websites include: www.worldbank.org; www.unctad.org; www.wto.org.

Check your understanding and progress at **www.hoddereducation.co.uk/myrevisionnotesdownloads**

Problems with special support measures include:
+ Not all poor countries are members of the WTO and application is a lengthy process.
+ Some low-income countries are not aware of the support available and therefore do not make full use of it.
+ There is concern from advanced economies that support of poor countries will lead to cheap imports flooding their markets.

The nature and role of transnational corporations

Transnational corporations (TNCs) dominate international trade – the top 500 TNCs account for more than 70 per cent of trade in goods and services. Most have a headquarters in advanced economies, especially the USA and Europe, and have branch plants in economies all over the world.

Advantages of multiple locations are that they can:
+ escape trade tariffs
+ find lowest-cost locations for production
+ reach foreign markets
+ exploit natural resources.

Spatial organisation

TNCs organise production to reduce costs, source raw materials and components at lowest-cost locations, and control key supplies. They have a common and recognisable global branding.

They outsource production and control processing at each stage of production.

They often exhibit three organisational levels:
+ Headquarters – generally in a city in an advanced country where they were first established.
+ Research and development – most likely at the headquarters or in another area within the same country.
+ Branch plants – located overseas, where costs can be minimised. These may vary according to the industrial sector. These are often manufacturing operations which take advantage of lower land, labour and material costs – offshoring. Outsourcing is a similar strategy where part of the business operation is subcontracted to another company in a country where costs are lower.

Primary sector activities are located where there are unexploited resources.

Secondary sector activities are located where:
+ labour costs are low
+ investment in education makes it easy to train workers
+ there is a work ethic of long hours and non-unionised labour
+ there are government incentives on offer, such as enterprise zones and tax breaks.

The service sector is more footloose – a company may factor in proximity to market in addition to the above. Language may also be a consideration, for example English speaking.

Trading and marketing patterns

There are two types of integration:
+ Vertical integration: the supply chain is owned entirely by the company.
+ Horizontal integration: the company diversifies its operation by expansion, merger or takeover.

Transnational corporations Companies operating in at least two countries with a headquarters in one country and other operations (branch plants) usually in a number of others.

Offshoring Relocating some part of a firm's activity to another country.

Outsourcing The process of subcontracting part of a firm's business to another company in order to save money.

Enterprise zone An area set up by the government to attract industry by the removal of taxes and restrictions to development.

Costs and benefits of TNCs

Table 7.4 sums up the costs and benefits of TNCs.

Table 7.4 Positive and negative impacts of TNCs for the host, the TNC itself and the country of origin

	For the host country	For the TNC	For country of origin (TNC base)
Benefits	Generates jobs and income Brings new technology Gives workers new skills Has a multiplier effect	Lower costs because of cheaper land and lower wages (fewer unions) Greater access to new resources and markets Fewer controls, such as environmental legislation	Cheaper goods Can specialise in financial services and R&D occupations
Problems	Poor working conditions Exploitation of resources Negative impacts on environment and local culture Economic leakages/repatriation of profits	Ethical issues such as the image of environmental damage or 'sweatshops' can be detrimental to their reputation Social and environmental conscience	Loss of manufacturing jobs Deindustrialisation Structural unemployment

Now test yourself

TESTED ◯

27 What are the disadvantages to a developing country of being overly reliant on the export of primary goods?

28 What are the advantages of export diversification?

29 Why do TNCs seek multiple locations?

30 Differentiate between outsourcing and offshoring.

Answers on p. 266

Example of a TNC and its impacts

Revision activity

Make detailed reference to a specified TNC and its impacts on those countries in which it operates.

For your chosen TNC, organise your revision notes into a diagram – you will probably need A3 paper to give you enough space.

Use the four bullet points below as headings to organise your notes. Remember to use bullet points; colour is also a way of aiding your revision – red for disadvantages, green for advantages, or different colours for different types of impact, for example blue for economic impacts.

The key to using detailed examples in an exam is to ensure that your points are specific to the example given – general statements that could be about any TNC, for example, will not gain the credit needed for a good answer. You must learn some place-specific facts.

Remember, the focus of this case study is impact.
+ Outline the nature of the industry.
+ Spatial organisation with reasons – this could be organised onto an annotated map.
+ Impacts on the country (countries) in which the TNC operates: economic/social/environmental/cultural.
+ Issues – for example, environmental consideration for location and policy.

Exam tip

In essays, provide a clear structure by subdividing impacts into:
+ social, economic, political and environmental impacts, or
+ impacts in countries where branch plants are located and impacts in the country of origin of the TNC.

Check your understanding and progress at **www.hoddereducation.co.uk/myrevisionnotesdownloads**

World trade in one food commodity or one manufacturing product

Analysis and assessment of the geographical consequences of global systems

The following have benefited from globalisation: newly industrialising countries (such as India), TNCs (Apple, Mondelez), international organisations (the IMF, World Bank, WTO), regional trading blocs (NAFTA, the EU). There are more specific economic, socio-cultural and environmental impacts of globalisation shown in Figures 7.9–7.11.

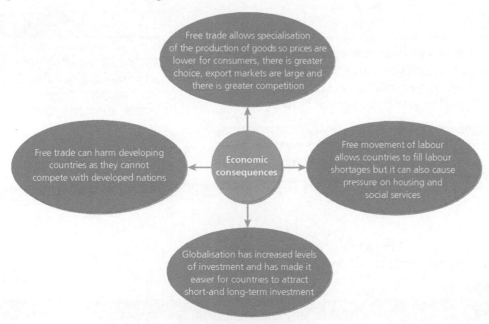

Figure 7.9 Economic consequences of globalisation

Figure 7.10 Socio-cultural consequences of globalisation

My Revision Notes: AQA A-level Geography Second Edition

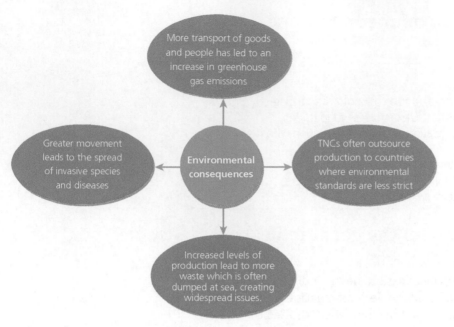

Figure 7.11 Environmental consequences of globalisation

Global governance

REVISED

Global governance usually works as a process of cooperative leadership whereby agreements that affect national governments are agreed. Recent focus of such agreements has been the environment, trade, poverty reduction, human rights, civil conflict, health issues and finance.

Global governance is seen as the most effective way to achieve sustainable development in an interdependent world. However, critics say that it undermines sovereignty of nations and marginalises poorer countries.

Global governance The way in which global affairs are managed through norms, laws, regulations and institutions.

Issues associated with attempts at global governance

Agencies such as those outlined in Table 7.5 can work to promote growth and stability but may also exacerbate inequalities and injustice.

Table 7.5 Global governance agencies

Agency	Role	Evaluation
UN	Aims to address global economic, social and environmental issues in a coordinated and collaborative manner Governance of the global commons – high seas, atmosphere, Antarctica and outer space Key involvement in climate change through the United Nations Framework Convention on Climate Change	Limited by size – difficult to gain consensus from almost 200 member states Developing countries often do not have the financial and technological resources to carry out international directives Concern that as Africa lags further behind economically, its influence on global governance will diminish
WHO	Main role is to direct and coordinate international health issues within the UN system World Health Assembly is the governing and decision-making body for WHO – 194 member states meet each year to set out policy	International cooperation remains primarily with governments but some are more powerful than others There is a question over the future role of countries like China, India, the USA and Russia, which all protect their sovereignty Criticism of the organisation being too bureaucratic and therefore inefficient, e.g. in Ebola and swine flu crises
WTO	Main role is to liberalise trade, provide a forum for negotiation of terms of trade A neutral body to aid the settling of trade disputes between nations	Positive terms of trade can lead to financial stability and the economic multiplier effect Many poor countries still have limited access to global markets Criticised as being ineffective in settling trade disputes

Check your understanding and progress at **www.hoddereducation.co.uk/myrevisionnotesdownloads**

Interactions at all scales

For global governance to be effective, clear communication is needed across all geographical scales, from global to local. Multi-scalar power operates through a range of scales to either encourage or prevent change. This involves a complex set of relationships from local to global level, as summarised in Figure 7.12.

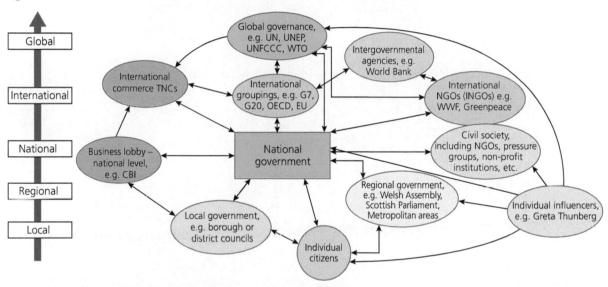

Figure 7.12 Multi-scalar web of interactions that influence global government

Attempts to achieve effective global governance include:

+ **Agenda 21:** the global blueprint for sustainable development. Agenda 21 action plans are intended to filter down from national governments to local communities; a 'top-down' approach aiming to encourage 'bottom-up' responses.
+ **UNFCCC Paris Climate Summit, December 2015:** a global agreement to combat climate change with a central aim to limit temperature increase to 1.5°C.
+ **Non-governmental organisations** such as Amnesty International, Greenpeace and Oxfam: these also have the scope to reach all scales and work closely with national governments and international organisations.

The global commons

REVISED ◯

Global commons are resource areas that lie outside the political reach of any one nation-state. The following are available for the use and benefit of all people:

+ the high seas (covered by the UN Convention on the Law of the Sea)
+ the atmosphere (covered by the UN Framework Convention on Climate Change)
+ Antarctica (covered by the Antarctic Treaty System)
+ outer space (covered by the 1979 Moon Treaty).

Due to current pressure on scarce resources the concept of the global commons is being contested. There is also pressure to maintain the right of all people to sustainable development and to protect the global commons.

Antarctica as a global common

The geography of Antarctica is outlined in Figure 7.13 while its role and some of the main issues affecting its vulnerability are summarised in the timeline in Figure 7.14.

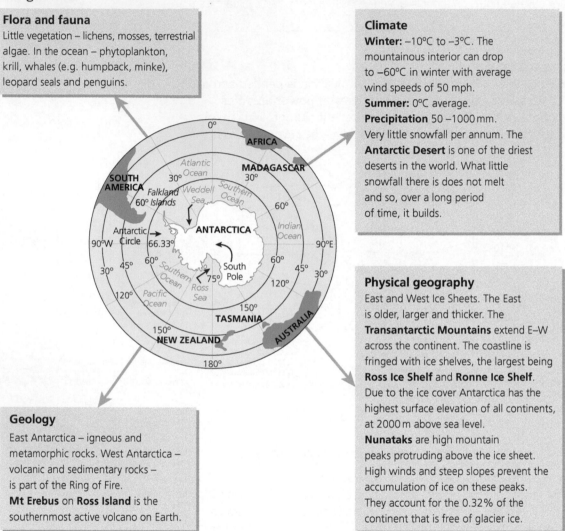

Flora and fauna
Little vegetation – lichens, mosses, terrestrial algae. In the ocean – phytoplankton, krill, whales (e.g. humpback, minke), leopard seals and penguins.

Climate
Winter: −10°C to −3°C. The mountainous interior can drop to −60°C in winter with average wind speeds of 50 mph.
Summer: 0°C average.
Precipitation 50 –1000 mm. Very little snowfall per annum. The **Antarctic Desert** is one of the driest deserts in the world. What little snowfall there is does not melt and so, over a long period of time, it builds.

Physical geography
East and West Ice Sheets. The East is older, larger and thicker. The **Transantarctic Mountains** extend E–W across the continent. The coastline is fringed with ice shelves, the largest being **Ross Ice Shelf** and **Ronne Ice Shelf**. Due to the ice cover Antarctica has the highest surface elevation of all continents, at 2000 m above sea level.
Nunataks are high mountain peaks protruding above the ice sheet. High winds and steep slopes prevent the accumulation of ice on these peaks. They account for the 0.32% of the continent that is free of glacier ice.

Geology
East Antarctica – igneous and metamorphic rocks. West Antarctica – volcanic and sedimentary rocks – is part of the Ring of Fire.
Mt Erebus on **Ross Island** is the southernmost active volcano on Earth.

Figure 7.13 The geography of Antarctica

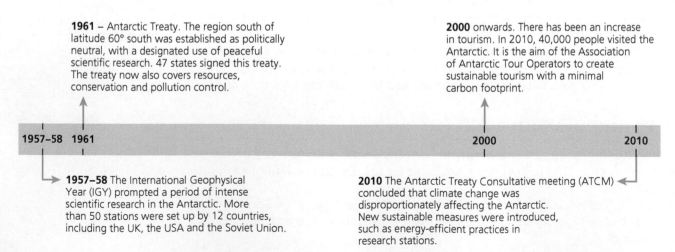

1961 – Antarctic Treaty. The region south of latitude 60° south was established as politically neutral, with a designated use of peaceful scientific research. 47 states signed this treaty. The treaty now also covers resources, conservation and pollution control.

2000 onwards. There has been an increase in tourism. In 2010, 40,000 people visited the Antarctic. It is the aim of the Association of Antarctic Tour Operators to create sustainable tourism with a minimal carbon footprint.

1957–58 The International Geophysical Year (IGY) prompted a period of intense scientific research in the Antarctic. More than 50 stations were set up by 12 countries, including the UK, the USA and the Soviet Union.

2010 The Antarctic Treaty Consultative meeting (ATCM) concluded that climate change was disproportionately affecting the Antarctic. New sustainable measures were introduced, such as energy-efficient practices in research stations.

Figure 7.14 Growing recognition of Antarctica's vulnerability

Check your understanding and progress at **www.hoddereducation.co.uk/myrevisionnotesdownloads**

Now test yourself TESTED ⬤

33 Why do we need to protect the global commons?

34 Describe the main features of the physical geography of Antarctica.

Answers on p. 267

The role and vulnerability of Antarctica

The role of Antarctica and some of the main issues affecting its vulnerability are summarised in the timeline shown in Figure 7.14.

Threats to Antarctica

Table 7.6 summarises the threats that Antarctica is facing.

Table 7.6 Threats to Antarctica

Threat	Issue (description)

Climate change	Climate change is affecting different parts of Antarctica in different ways. According to the IPCC, the Antarctic peninsula is particularly sensitive to small rises in annual average temperature. Changes include the following:
	 ✦ Temperatures in the Southern Ocean to the west of Antarctica have increased with the following effects:
	   ✦ changed distribution of penguin colonies
	   ✦ decline in Antarctic krill
	   ✦ retreat of glaciers and ice shelves fringing the peninsula.
	 ✦ Sea ice to the east is increasing due to:
	   ✦ increased westerly winds around the Southern Ocean driving seas northwards
	   ✦ more rain and snow resulting from climate change, layering the Southern Ocean with cooler, denser air
	   ✦ greater melting of continental land ice creating more floating icebergs which contribute to sea ice formation.
	 ✦ Ocean acidification. CO_2 in the atmosphere creates carbonic acid, which makes the ocean less alkaline. The loss of carbonate ions may lead to waters becoming corrosive to unprotected shells and the loss of these organisms will disrupt foodwebs.

Fishing and whaling	1986: globally, most whaling stopped due to the action of the International Whaling Commission (IWC). 1994: the IWC established the Southern Ocean Whaling Sanctuary, which banned commercial whaling. Fishing has replaced whaling in the area. Commercial fishing is becoming a significant threat to the Southern Ocean and Antarctica, with high levels of over-fishing, particularly of krill, by Russia and Japan. Ships also dump waste into the ocean and destroy marine habitats.
Search for mineral resources	Mining is banned by the Antarctic Treaty. Demand for resources puts pressure on the mineral resources on the continent, which include gold, silver, lead and zinc.

Tourism and scientific research	The continent is only populated by scientists at a small number of research stations. Scientists are well briefed in appropriate care for the environment, but inevitable damage is caused by fuel storage and the disposal of waste.
	
	 Tourism is of three types: camping trips, ship-boarding sites, over-flights.
	 ✦ Antarctica is a niche destination due to the wildlife (seals and penguins in particular), remoteness and ice coverage.
	 ✦ Tourism is limited to mid-November to March.
	 ✦ There is a strict code of conduct, health, and safety. Numbers of tourists continue to rise and in the 2018–19 season in excess of 50,000 people visited Antarctica.
	 ✦ Tourism is governed by the International Association of Antarctica Tour Operators (IAATO).
	 ✦ The impact of tourism is monitored by the Scott Polar Research Institute. Despite positives in the way that the tourism is run and controlled – for example, widely accepted guidelines, very low litter levels attributed to tourists, only 5% of 200 landing sites for tourists showing any wear and tear – there are concerns:
	   ✦ Summer tourists coincide with peak wildlife breeding.
	   ✦ Land-based installations are clustered in the few ice-free locations.
	   ✦ There is a threat of land-based tourism.
	   ✦ The demands for fresh water are difficult to meet.
	   ✦ There is evidence of the over-flying causing stress to breeding colonies.

Critical appraisal of the developing governance of Antarctica

Resilience, mitigation and adaptation are monitored by the Scientific Committee on Antarctic Research (SCAR).

Climate change is the major threat to the Antarctic – see Figure 7.15. The following are threatening the resilience of species and their ability to adapt:

✦ increasing sea temperatures
✦ ocean acidification
✦ expanding sea ice in some areas
✦ loss of sea and land ice in other areas
✦ high intensities of ultraviolet radiation.

Making links

See Chapter 1 for an explanation of the impacts of climate change.

> **Resilience** The extent to which an area can recover from the impact of something negative, for example an oil spill or over-fishing.
>
> **Mitigation** The action of reducing the severity of something.
>
> **Adaptation** A change that allows survival or coping mechanisms.

Exam tip

Be prepared to evaluate the extent of each of the threats in Table 7.6 in light of the opinion that climate change poses the major threat.

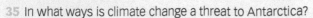

Figure 7.15 Linking adaptation and mitigation in an integrated framework of climate change

A timeline of the key developments in the governance of Antarctica is shown in Figure 7.16.

Revision activity

For each of the threats in Table 7.6, make revision notes on strategies for
a) **mitigation** (the action of reducing the severity of something), b) **resilience** (the extent to which an area can recover from the impact of something negative) and c) **adaptation** (change that allows coping mechanisms). For example:

Tourism
✦ Mitigation – Antarctic tourism is well governed, with strict rules set out by the IAATO.

Check your understanding and progress at **www.hoddereducation.co.uk/myrevisionnotesdownloads**

Pre-WW2	Seven countries make territorial claims on sections of Antarctica – based on exploration history (UK, Norway) or being supra-adjacent nations (Chile, Argentina, France, Australia and New Zealand) – these claims were never agreed but they are still known as 'claimant countries'.

Recognising that disputed status needed to be addressed, scientists appeal to UN to establish an **International Geophysical Year (IGY)** to promote scientific cooperation. Scientists from 12 countries with already established scientific research stations worked cooperatively on a multinational research programme, which was a great success.

1950s: 1957–58 (IGY)

1959–1961 The success of the IGY led to the 12 nations involved (the seven 'claimants' plus Belgium, Japan, South Africa, Soviet Union and USA) resolving the issue by signing the **Antarctic Treaty** (AT). Declared that Antarctica and the surrounding ocean up to 60°S would be a zone of peace and cooperation, guaranteeing free access and research rights to all countries.

Increased pressure and demand for opening up of Antarctica's resources; new states including China and India had joined the AT. Members began to negotiate new regimes for fishing rights and also considered a Mineral Convention to allow mining. Countries in the Global South complained to UN General Council that a small group were trying to exploit resources in an area of 'common heritage', which led to further protection and development of the wider **Antarctic Treaty System (ATS)** with added agreements.

1980s

1991–98 The **Protocol on Environmental Protection to the Antarctic Treaty** (Madrid Protocol) was signed (1991) and ratified (1998), giving further protection to Antarctica, particularly from mining, until 2048.

The **Southern Ocean Whale Sanctuary** was designated by the International Whaling Commission. 23 members supported it, but Japan opposed it.

1994

2017 The world's largest **Marine Protected Area (MPA)** was designated in the Ross Sea by the **Commission for the Conservation of Antarctic Marine Living Resources (CCAMLR)**. 72% of this is fully protected.

Japan leaves the IWC. China and Russia are both in favour of greater exploitation of fishing resources and, in 2018, together with Norway, blocked plans for an even larger MPA in the Weddell Sea. Other parties want to see more protected areas.

2018–present

Figure 7.16 Timeline of key development in the governance of Antarctica

---

Now test yourself TESTED ○

38 Explain the process of ocean acidification.

39 Why do some countries back greater exploitation and exploration of resources, for example fish and minerals, in Antarctica?

Answers on p. 267

The role of NGOs in the protection of Antarctica

Antarctic and Southern Ocean Coalition (ASOC)

This is a coalition of more than 30 NGOs, including Friends of the Earth, the Worldwide Fund for Nature and Greenpeace. Its main aims are to:

+ convince governments to conclude negotiations of the 'ecosystem as a whole' treaty on fishing
+ prevent exploitation of oil, gas and minerals
+ open up the Antarctic Treaty System to include NGOs.

The ASOC has had success when:

+ a precautionary ecosystem approach was embedded into the Antarctic Treaty
+ the Minerals Convention was blocked
+ it was instrumental in the development of the 1991 Madrid Protocol.

ASOC currently has campaigns in:

+ negotiating a Polar Code covering vessels operating in the Southern Ocean
+ establishing a network of marine reserves
+ managing Southern Ocean fisheries, including krill sustainability
+ strengthening the Southern Ocean Whale Sanctuary.

Exam tip

Note the wording of the specification for this section is 'critical appraisal'. If you are asked to evaluate or assess in an exam question this means that you have to do more than just name the governing bodies involved and describe what they do – you need to express an opinion on how effective they are in terms of the protection of Antarctica.

Analysis and assessment of global governance

Once you have revised the section on global governance, attempt the following practice essay.

'Tourism in Antarctica should cease.' How far do you agree with this statement?

Consider the following:
+ What are the costs and benefits of tourism in Antarctica?
+ Think about the following categories: cultural/social/economic and environmental.
+ To what extent is exposure to tourists important to global understanding of Antarctica's importance as a global common?
+ How effective is current governance of tourism in Antarctica?

Remember to plan your essay. Use place-specific facts. Present a balance – in what ways do you agree with the statement and in what ways do you disagree?

Exam tip

There are 20 marks allocated for essays. You are given 40 lines of exam script as a guide.

Globalisation critique

REVISED

There is a range of views on globalisation (Table 7.7).

Table 7.7 Views on globalisation

	Views on globalisation
Hyperglobalists	+ In effect, they 'support' globalisation. + They see the nation state as no longer important. + They view a new geographical era in which there is a single global market supported by extensive and open networks and flows of goods, information, people and finance.
Sceptics	+ They hold the view that globalisation is nothing new and that the world has always been integrated. + They are sceptical of the free movement of goods aspect of the hyperglobalist view as many countries adopt protectionist measures. + China, India and the USA have achieved their growth through government policies and upholding their sovereignty. + They also believe that globalisation marginalises the poor.
Transformationalists	+ They hold a view in between the above. + They accept the process of increasing globalisation but think that the role of governments is changing rather than being overtaken by group decision making. + They also acknowledge the time–space compression through extensive new networks and flows.

Summary of benefits and costs of globalisation

Using your own notes or from further research, create a table to summarise the costs and the benefits of globalisation. Think about the following:
+ how globalisation has positively impacted growth and economic development
+ the negative aspects of inequality
+ the environmental costs of globalisation
+ the conflicts that globalisation has caused, for example the social and cultural conflict caused by international migration.

Review this chapter by conducting your own critique of the globalisation process.
+ **S**trengths   + **W**eaknesses   + **O**pportunities   + **C**osts

Make revision notes under each of these headings in the form of a table or diagram.

Making links

The globalisation process will impact climate change and the functioning of the carbon cycle. See Chapter 1 page 29 for more.

Exam practice

1 Explain two features of the flows shown in Figure 7.8 on page 152 – patterns of intra-regional and inter-regional world trade. [4]

2 How have transport systems influenced globalisation? [4]

Answers and quick quizzes online

3 Explain why less developed countries have difficulty accessing global markets. [6]

4 Analyse the socio-cultural impacts resulting from globalisation. [6]

Exam skills

Opportunities to practise geographical skills within this topic include:

+ observation skills in fieldwork
+ collection, manipulation, presentation and analysis of primary and secondary fieldwork data – including quantitative and qualitative sources
+ cartographic skills – annotation and analysis of a range of maps presenting flows (for example of people and finance) and patterns (for example international trade)

+ analysis of geo-located data and digital imagery, for example Antarctica
+ interpretation of a range of graphs – bar charts, divided bar charts, histograms, Lorenz curve (representing inequalities for example), line graphs
+ the use of statistical tests to manipulate data, for example Spearman's rank correlation coefficient to examine the relationship between a range of factors such as GDP, FDI and economic growth.

Summary

+ Globalisation is a complex process with social, political, cultural, economic and environmental dimensions.
+ 'Flows' is a useful umbrella term to explain the process of globalisation. It includes flows of people, goods, capital and ideas.
+ Different factors have led the drive towards globalisation: financial, transport innovation, security, communication, information systems and trade.
+ Global affairs are managed by international political organisations. Have a clear understanding of the role and the criticisms of the major ones: World Bank, the IMF, the WTO and the IPCC.
+ A number of systems support the economic, political, social and environmental interdependence that has resulted from globalisation.
+ Increased global interdependence raises issues of unequal flows and unequal power.

+ International trade is of particular importance as it drives economic development. It is important to understand the patterns, trends, relationships and accessibility of markets.
+ Transnational corporations are key players in the globalisation process. The features and impacts of TNCs should be studied through a detailed example.
+ The Antarctic is one of the global commons. Develop a clear overview of its geography, role and vulnerability.
+ The governance of the Antarctic faces many challenges now and in the future.
+ Develop your own critical evaluation of the globalisation process and be aware of the costs and benefits involved.

8 Changing places

The nature and importance of places

REVISED

The concept of place and the importance of place in human life and experience

Place is a key term in geography. It can be seen as a **location** on a map or more broadly as a **description** of human and physical characteristics. There are three key concepts of place (see Figure 8.1):

+ location
+ locale
+ sense of place.

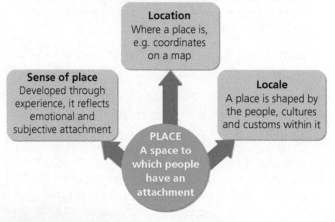

Figure 8.1 The different aspects of place

Exam tip

Make sure you can illustrate the meaning and concept of place by referring to located examples; this can be based on your local area or an experienced place.

Revision activity

Apply the different aspects of place to a local place with which you are familiar. Summarise a brief description of its location, locale and your lived experience of that place.

Theoretical approaches to place

There have been frequent discussions among academic geographers on **the meaning of place**. Table 8.1 summarises the main theoretical approaches to explaining the concept.

Table 8.1 Key theoretical views on the meaning of place

Approach	Description	Example
Descriptive approach	The world is made up of a set of places, each of which can be studied through a description of its physical and human characteristics.	A descriptive geography of your local area
Social constructionist approach	Places are the product of social processes occurring at a particular time and place and it is this that gives meaning to a place.	(see revision activity)
Phenomenological approach – Yi-Fu Tuan and Edward Relph	Places are defined through human experiences. It is the connection between place and person that transforms unknown spaces into familiar places.	(see revision activity)

Exam tip

Develop a critical evaluation of the usefulness of different theoretical views on the meaning of place.

Revision activity

Complete Table 8.1 with a located example for each approach.

The importance of place in human life and experience

Place is important for the following reasons:
+ It gives people a personal identity.
+ Marketing and positive image building will attract visitors.
+ People will have personal preferences of the type of place in which they would like to live (rural, urban, coastal) or holiday.

People have a **lived experience** of place that can impact on:
+ identity – identification with place exists on a local, regional and national scale
+ sense of belonging
+ well-being.

Globalisation and localisation of place

Globalisation is the growing interdependence of countries through:
+ increasing global transactions of goods and services
+ increasing flows of information, labour and capital
+ the widespread transfer of technology.

Many believe that this has given rise to a new geographical era of **'placelessness'**, where global capitalism has eroded local culture and localised identities and produced identical or homogenised places.

An example is the global spread of retail chains and TNCs such as McDonald's, Costa Coffee and Hilton Hotels, meaning that city centres across the world have common elements.

Glocalisation is a response to globalisation. This centres on the promotion of **local** goods and services and the adaptation of global products to the specific locality in an effort to regain local cultures and identities.

Examples include McBurritos in Mexico, and in Totnes in Devon, where there is support by the local council for independent coffee outlets run by locals rather than chains such as Starbucks and Costa. (However, in 2019 a coffee shop from a different national chain opened a branch in Totnes.)

Making links

See Chapter 7 page 143 for more on globalisation and its impacts on places.

Homogenise To make a place uniform or similar.

Now test yourself

1 What are the three key concepts of place?
2 Explain what is meant by homogenised places.
3 What is glocalisation?

Answers on p. 267

TESTED ⬤

Revision activity

Complete Figure 8.2 by answering the questions in the middle boxes to outline how places are becoming more homogenised (to make a place uniform or similar) as a result of globalisation and how places are attempting to regain a sense of local identity as a result of glocalisation.

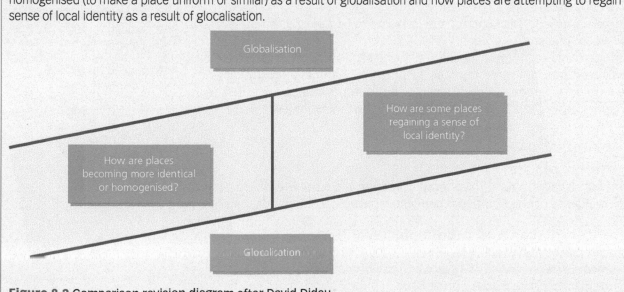

Globalisation

How are some places regaining a sense of local identity?

How are places becoming more identical or homogenised?

Glocalisation

Figure 8.2 Comparison revision diagram after David Didau

The importance of place

The place-making movement of recent years identifies three key factors of the importance of place and how it defines people (see Figure 8.3):

✦ identity ✦ belonging ✦ well-being.

Well-being
What makes a good place to live?

Promotion of well-being through:
- sociability – neighbourly, friendly, welcoming
- access – connected, access to resources and services
- activities – social, community
- image – clean, safe, attractive, stable.

Identity
How is place important to you?
How does place define your identity?

Place can be critical to the construction of a person's identity. It is enhanced by knowledge and experience and is evident at different scales:
- local – emotional ownership
- regional – loyalty to a region
- national – patriotism.

The importance of place

Belonging
What does it mean to belong to a place?
What influences the feeling of belonging?

Sense of belonging fostered through:
- community spirit
- inclusivity regardless of age, gender, socio-economic status, religion, race.

Figure 8.3 The importance of place

Insider and outsider perspectives on place

Different people perceive places in different ways. This may in turn give rise to a sense of being an **insider** or an **outsider**.

For any one place some people may feel a strong sense of attachment and belonging while others feel detached. Factors affecting this sense of attachment include race, ethnicity, age, religion, socio-economic status as well as types of experience.

✦ Refugees often feel 'out of place' as they have been forced to flee their native country.
✦ An urban area, such as London, will be home to people of different levels of socio-economic status, whereby:
 ✦ a high wage earner may develop a strong sense of 'insider' as they have full economic access to services, facilities, jobs and entertainment
 ✦ a homeless person or one of low economic status will have limited opportunities, which restricts their ability to feel included.

Categories of place

Near and far places

Near and far places can refer to geographical distance, a socio-economic gap between residents or the emotional connection people feel towards a place.

Experienced places and media places

Experienced places and media places refer to the difference between the reality of people's experience of a place and the often more selective media image – for instance, the media portrayal of:

✦ the rural idyll as opposed to the negative portrayal of inner-city areas
✦ places for the attraction of investments or tourism.

Check your understanding and progress at **www.hoddereducation.co.uk/myrevisionnotesdownloads**

Making links

See Chapter 7 for the subsystems of globalisation and how social globalisation is affecting images of places and their people.

Revision activity

Choose two contrasting places, for example rural/urban, large/small, familiar/unfamiliar, UK/international, mountainous/coastal, and compare media images of each place. How are they characterised and described in the media and what sorts of stereotypes does this create?

Factors contributing to the character of places

Character refers to the human and physical features of a place that give it its unique identity. In general, there is a range of factors that affect the character of places (see Figure 8.4).

Physical geography such as relief, altitude, aspect, drainage, soil and rock type

Socio-economic factors such as employment opportunities, amenities, educational attainment and opportunities, income, health, crime rates, local clubs and societies

Cultural factors such as heritage, religion, language

Demographic factors: population size and structure (age and gender), ethnicity

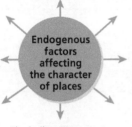

Endogenous factors affecting the character of places

Mobility of the population for work and leisure pursuits

Location: urban or rural, proximity to other settlements, main roads and physical features such as rivers, the coast, etc.

Political factors such as the role and strength of local councils and/or resident groups

The built environment: land use, age and type of housing, building density, building materials

Figure 8.4 Endogenous factors affecting the character of places

The impact of international migration. People from all around the world have settled in the UK, sometimes forming **diasporas** and definitely creating a more multicultural society

Links to or influences from other places and distances from and route ways to other places and the accessibility of place

Deindustrialisation, unemployment, economic restructuring and urban decline of former traditional industrial cities as manufacturing moves overseas

Exogenous factors

'Newcomers' arriving in an area could cause conflict or change the character or community of a place. New housing estates may be built in villages, second homes purchased in seaside resorts or inner cities may be gentrified

The increasing mobility of people increases the links between places

Mining, steel and shipbuilding towns having to adapt to the challenges posed by globalisation

Figure 8.5 Exogenous factors affecting the character of places

Endogenous and exogenous factors

Specifically, there are **endogenous factors** of internal origin – location and physical geography – as well as social, economic and demographic factors (see Figure 8.4) and **exogenous factors** of external origin – the shifting flows of people, capital and resources, including links with and influences of other places (see Figure 8.5).

Now test yourself   TESTED

7   Define exogenous and endogenous factors in the context of place studies.
8   What is **deindustrialisation** and how can it impact the character of a place?

Answers on p. 267

Exam tip

When revising exogenous and endogenous factors affecting the character of places, think about the impact they will have on a specific place. For example, unemployment affects people's well-being and will also present social challenges associated with low incomes and poverty.

Revision activity

Make a copy of Figures 8.4 and 8.5 but at the centre include the name of your local place study. For each stem add place-specific details relating to your local place study (e.g. the physical geography of your place, the nature and character of the built environment).

Changing places   REVISED

Figure 8.6 shows the structure of the remaining sections of this unit, which you will have studied within the context of two contrasting places:

+ a **local** place – local to your home or study area
+ a **contrasting** place – likely to be distant, either within the same country or in a different country.

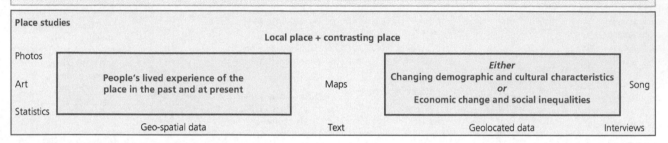

> **Changing places**
> **Key concept:** People's engagement with places. How are places known, experienced, characterised and impacted upon by change?
>
> **Study must be embedded in a local and contrasting (distant) place.**
>
> **Relationships and connections**
> - The demographic, socio-economic and cultural characteristics of places are shaped by shifting flows of people, resources, money, investment and ideas. These flows operate at scales from local to global.
> - The impacts of external forces operating at different scales from local to global. These include: government policies; decisions of multinational corporations; the impacts of global institutions.
> - Past and present connections shape places and embed them in the regional, national, international and global contexts.
> The impacts of relationships and connections on people and place focuses on **either**:
> **Changing demographic and cultural characteristics**
>   **or**
> **Economic change and social inequalities**
>
> **Meaning and representation**
> - Meaning relates to perceptions of place.
> - Representation is how a place is portrayed in society.
> - Both change over time. Meaning and representation are shaped through people's lived experience of a place.
>
> Changing Places
>
> **Themes studied**
> - How people perceive, engage with and form attachments to place.
> - How external agencies influence place meaning.
> - How places are represented in a variety of forms.
> - The influence of past and present processes.

Place studies

Local place + contrasting place

Photos / Art / Statistics

People's lived experience of the place in the past and at present — Geo-spatial data

Maps — Text

Either Changing demographic and cultural characteristics *or* Economic change and social inequalities — Geolocated data

Song — Interviews

Figure 8.6 Structure of the Changing places unit

Relationships and connection

Places change over time due to changes in:

- people
- resources
- money and investment
- ideas
- conflict
- migration
- natural disasters.

The focus of this section is on the impact of **relationships and connections** on people and place. Table 8.2 shows some examples of **agents of change** and their **impact**.

> **Now test yourself**
>
> 9 Name two ways in which people's perception of place can be formed.
>
> 10 Give two reasons why a government might adopt strategies to change people's perception of **place**.
>
> **Answers on p. 267**
>
> TESTED ◯

Table 8.2 Some impacts of external forces on place

Agent of change	Example	Impact
Government policies	Regeneration schemes and financial incentives for industries, such as subsidies, tax breaks and enterprise zones	These can attract businesses to places and stimulate a positive multiplier effect.
The decisions of multinational corporations	In 2010, Mondelez International closed the Cadbury factory near Bristol and moved production to Poland. In 2016, Tata Steel announced UK job cuts in response to difficult global market conditions.	Job losses for employees. Factories converted into housing. Major job losses at Port Talbot, Hartlepool and Corby – all highly dependent on the steel industry
The impacts of international or global institutions (for example, IMF, World Bank, UN, WHO)	In 2015, the World Bank was running 15 development projects in Haiti. Millennium Development Goals	Post-earthquake reconstruction of both homes and communities. Varied level of success around the world

Check your understanding and progress at www.hoddereducation.co.uk/myrevisionnotesdownloads

Meaning and representation of place

Meaning and representation is defined in Figure 8.6.

People's perceptions of place and how places are portrayed in society can both change over time.

Perception of place can have two sources, essentially from direct experience or from relayed information from other sources:
+ **Meaning** relates to individual and collective perceptions of place.
+ **Representation** is how a place is portrayed in society.

Both meaning and representation change over time.

Sense of place
A developed **'sense of place'** is the result of lived experience in a location – developing a sense of place is important to an individual's personal identity.

Perception of place
A **perception** of place is developed through what people have heard or read about a place.

Influence on and manipulation of the perception of place

Often agents of change, such as government (local and national), corporate bodies, tourist organisations and local community groups, will attempt to manage or direct the perception of a place.

Governments often do this to attract people and investment to an area. The following means can be used and can probably be identified in your place studies:
+ Place marketing includes:
    + advertising campaigns – online and through hard-copy newspapers, fliers, magazine articles
    + events – food festivals, Christmas markets, music events, cultural events.
+ **Rebranding** is often used to dispel negative perceptions. Rebranding aims to give a place a new, more positive identity at the local, national and international levels.
+ **Re-imaging** is related to both rebranding and regeneration and involves a marketing/public relations exercise to promote a modern, fresh and positive image of a place.
+ **Regeneration** is a longer-term process often aimed initially at economic regeneration, which through the multiplier effect will bring further social and physical improvements.

Figure 8.7 provides a summary of the strategies outlined above.

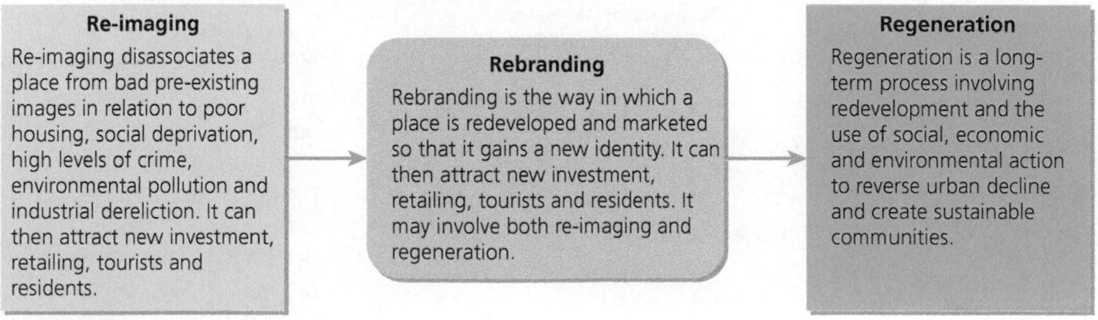

> **Exam tip**
>
> **Change over time** is a much-used concept in geographical studies. Be prepared to explain how places you have studied have changed over time in terms of both meaning and representation.

> **Exam tips**
>
> Image is a strong determinant of human behaviour. People's perception of place is formed by positive images, particularly when they have no lived experience of that place.
>
> Do not see the components of urban rebranding in isolation – frequently, re-imaging and regeneration lead to rebranding.

Re-imaging

Re-imaging disassociates a place from bad pre-existing images in relation to poor housing, social deprivation, high levels of crime, environmental pollution and industrial dereliction. It can then attract new investment, retailing, tourists and residents.

Rebranding

Rebranding is the way in which a place is redeveloped and marketed so that it gains a new identity. It can then attract new investment, retailing, tourists and residents. It may involve both re-imaging and regeneration.

Regeneration

Regeneration is a long-term process involving redevelopment and the use of social, economic and environmental action to reverse urban decline and create sustainable communities.

Figure 8.7 The components of urban rebranding

> **Exam tip**
>
> Develop a critical evaluation of attempts to manage place perception in your place studies. Column three of the revision activity on page 171 shows what is meant by **'evaluation'**. It involves a balanced assessment of positive and negative factors.

169

Revision activity

Based on a place you have studied, make revision notes in the form of a table which summarises the:

a) reasons for managing place perception and representation

b) attempts made to manage place representation

c) extent to which these measures have been successful.

For example, based on Amsterdam:

Reasons for management	Strategies to influence representation	Evaluation of success
Competition from other cities	Rebranding and use of the 'I amsterdam' logo	Highly successful – the logo emerged as the city's most photographed item. The image spread globally through social media. But – some think it has been too successful and has led to over-tourism.

Now test yourself

11 What do the following terms mean:
   a) rebranding,
   b) reimaging,
   c) regeneration?

12 Name two types of place that might adopt rebranding strategies.

13 Give three stakeholders involved in rebranding strategies.

Answers on p. 267

TESTED

Place studies

REVISED

For this chapter you will have studied the changing character of **a local place** and **a contrasting place**. The focus will be on:

+ people's lived experiences, past and present
+ either changing demographic and cultural characteristics or economic change and social inequalities.

Revision activity

Figure 8.8 sets out a grid showing how you could summarise your revision notes for your two place studies.

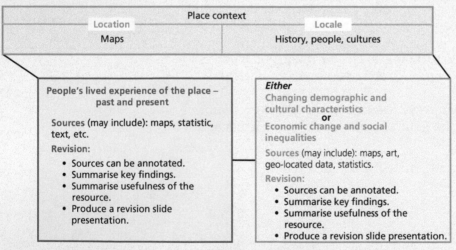

Figure 8.8 Possible structure for place study revision notes

Possible sources of information include:

+ statistics and census data – to illustrate demographic, social and economic characteristics
+ maps – showing changes over time
+ geolocated data
+ geospatial data
+ photographs – to illustrate representations of place
+ various forms of text – poetry, novels, newspapers
+ audio-visual material
+ art
+ oral sources – interviews, music.

Exam tips

It is important to develop a critical evaluation of the usefulness of a range of qualitative and quantitative resources and revise these as a bullet point list, along with the findings from each resource.

Interviews provide a wealth of information on people's lived experiences of place. Make sure that questions are clear. Summarise key findings for revision.

Revision activity

There will be several key terms in this topic that you have not come across before in your study of geography. Make a list with definitions and include place, locale, location, perception of place, sense of place, endogenous factors, exogenous factors, representations of place.

Check your understanding and progress at **www.hoddereducation.co.uk/myrevisionnotesdownloads**

Exam practice

1 Compare and contrast two theoretical views on the meaning of place. [4]

2 Explain how place can define a person's identity. [4]

3 Compare the place profile of Lympstone (East Devon) and Toxteth (Liverpool) by analysing the data shown in Table 8.3, Table 8.4 and Figure 8.9. [6]

4 Explain how physical geography has influenced the identity of a place you have studied. [6]

5 With reference to a named place you have studied, assess the impact of local government policies on place representation. [6]

Table 8.3 Some key demographic characteristics of Lympstone and Toxteth (from 2011 census)

Variable	Lympstone	Toxteth	England
Population density (persons per hectare)	16.6	87.8	4.1
0–15 years old	19.8	13.9	18.9
16–64 years old	55.6	75.5	64.8
65 years old and over	24.6	10.6	16.3

Table 8.4 Some socio-economic characteristics of Lympstone and Toxteth

Variable	Lympstone	Toxteth	England
Average household size (number of people)	2.3	1.9	2.4
Owner-occupiers	66.1	24.0	64.5
Rent from social landlord	12.2	34.2	17.6
Rent from private landlord	20.6	39.3	16.7
Car availability (% with no access to a car or van)	12.7	54.4	25.6
% people with bad or very bad health	4.1	9.4	5.6
% aged 16 or over with no formal qualifications	14.1	27.2	22.5

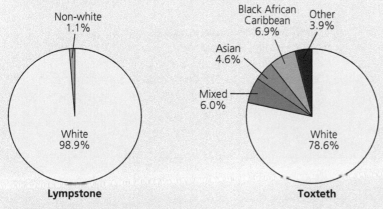

Figure 8.9 Ethnic composition of Lympstone and Toxteth

Answers and quick quizzes online

171

Exam skills

Opportunities to practise geographical skills within this topic include:
+ representation of place and the use of qualitative and quantitative sources
+ places can be represented in a range of forms
+ you can use a variety of quantitative (data) and qualitative (non-numerical) sources to investigate and present your place studies (Table 8.5).

Table 8.5 Quantitative and qualitative skills and sources for investigating and presenting place studies

Quantitative skills and sources	Qualitative skills and sources
Use of statistics: + Census data + Office for National Statistics (ONS) data These are the main sources for data on demographic characteristics of places. **Statistical techniques:** Such data can be manipulated using a variety of statistical techniques, including: + ratios, fractions and percentages + measures of central tendency (e.g. frequency distributions) + measures of correlation (e.g. Spearman's rank correlation coefficient). **Graphical presentations:** You can present data in the form of a range of graphical presentations: + histograms + bar charts + divided bar charts + population pyramids + line graphs + Lorenz curve graphs. Annotations to charts and graphs provide analysis for revision.	**Interpretation of a range of maps:** + Ordnance Survey (OS) maps + Goad maps + Google maps + maps provided and adapted from Geographical Information Systems (GIS). **Ethnographic research:** This focuses on what people say and do and may include: + interviews + surveys and questionnaires. You can annotate written transcripts of interviews with analytical comments. The results of surveys and questionnaires can also be analysed quantitatively with statistical techniques. **Other qualitative sources include:** + photographs + written text (e.g. novels, poetry, newspaper reports, magazine articles, promotional materials) + music + TV + film + art – paintings, sculpture + the internet – blogs, sources of statistics, facts, opinions.

Summary

+ Location, locale and sense of place combine to form the concept of place.
+ Develop a critical evaluation of the theoretical views on what is meant by a sense of place.
+ Categories of place include near and far, and experienced and media places.
+ A range of factors contributes to the character of places. These include endogenous and exogenous factors.
+ Relationships, connections, meaning and representation of place should be understood in the context of a local and a contrasting place.
+ For your chosen case studies, changing demographic and cultural characteristics or economic change and social inequalities should be understood in both time and scale (regional, national, international and global) contexts.
+ Understand how people's perception of place is formed and the influence of different agencies on this perception.
+ Make detailed revision notes with place-specific facts on a local and contrasting example to illustrate people's lived experience of place, past and present.
+ Case studies will be based on a range of research resources, which should be critically evaluated in terms of their usefulness.

Check your understanding and progress at **www.hoddereducation.co.uk/myrevisionnotesdownloads**

9 Contemporary urban environments

Urbanisation

REVISED ●

Urbanisation and its importance in human affairs

+ Urbanisation is the process by which an increasing proportion of a country's population lives in towns and cities.
+ Its visible outcome is a rise in the proportion of a population living in urban areas.
+ Urbanisation rates are usually expressed as a percentage.
+ The process is reflected in:
  + a shift in the economy away from primary activities towards manufacturing and services
  + more people living in built-up areas
  + the physical spread of built-up areas.

The importance of this process in human affairs involves the following:

+ By 2050, it is expected that 66 per cent of the global population will live in urban areas and most of this number will be concentrated in Asia and Africa. The challenge will be to provide these growing urban areas with the basic resources that people need and to manage the strain on the environment.
+ Urban areas are absorbing the pressure of global migration patterns and also the movement of people from rural to urban areas within a country (as agricultural practices are transformed).
+ The sustainable growth of cities will be a major challenge of the twenty-first century.

Global patterns since 1945

The world's urban population grew from 746 million in 1950 to 3.9 billion in 2014.

1950–1990

+ Prior to 1950, the majority of urbanisation occurred in high–income countries. Rapid urbanisation took place during the period of industrialisation in Europe and North America in the nineteenth and early twentieth centuries. Many people moved from rural to urban areas to get jobs in the rapidly expanding industries in many large towns and cities.
+ Since 1950, urbanisation has slowed in many high-income countries and now some of the biggest cities are losing population as people move away from the city to rural environments.
+ Since 1950, the most rapid growth in urbanisation has occurred in low-income countries in South America, Africa and Asia.
+ Between 1950, and 1990, the urban population living in low-income countries doubled. In high-income countries the increase was less than half.
+ There are three main causes of urbanisation in low-income countries since 1950:
  + Rural to urban migration is happening on a massive scale due to population pressure and lack of resources in rural areas.
  + People living in rural areas are 'pulled' to the city. Often, they believe that the standard of living in urban areas will be much better than in rural areas.
  + There is a natural increase caused by a decrease in death rates while birth rates remain high.
+ By 1990, there were megacities.

> **Industrialisation** The process by which an economy is transformed from primarily agricultural to one based on the manufacturing of goods.
>
> **Megacity** A city with a population in excess of 10 million people.

173

1990 to present day

+ Economic contraction led to population losses in some American cities.
+ There was an increase in the number of megacities to 30.
+ Between 2010 and 2014, some of the largest growth rates were in China, India, and Nigeria.
+ In 2015, Tokyo was the world's largest city, closely followed by Delhi, Mumbai and Shanghai.
+ In 2014, the most urbanised regions were North America (82 per cent), Latin America and the Caribbean (81 per cent) and Europe (74 per cent).
+ Presently, the fastest-growing urban areas are in Asia and Africa.
+ An example of economic contraction is Detroit, USA.
+ An example of emigration is Poznan, Poland.

> **Now test yourself** TESTED
>
> 1 State three indicators of rising levels of urbanisation.
> 2 Why will the growth of cities present a major challenge for the twenty-first century?
> 3 Outline two causes of urbanisation in low-income countries since 1950.
>
> **Answers on p. 267**

> **Exam tip**
>
> 'Push and pull factors' is a useful way of explaining the causes of these processes. For example, suburbanisation – lack of space in central areas and the attraction of space in the suburbs.

Predictions for present day onwards

+ The UN predicts that, by 2030, 60 per cent of the world's population will live in urban environments and by 2050 that figure will have reached 66 per cent.
+ It is predicted that there will be 37 megacities by 2025.
+ Brazil is expected to experience rapid increases in its urban population by 2030.
+ By 2050, the World Bank predicts that China, India and Nigeria will account for 35% of the growth of the world's urban population.

Urbanisation, suburbanisation, counterurbanisation, urban resurgence

Several key processes have been associated with the patterns in urbanisation identified above, particularly in more developed economies since 1945. Figure 9.1 summarises these processes and the sequence in which they occur.

Urbanisation
Increasing proportion of total population living in towns and cities. Population growth in urban areas coming from immigration (from rural areas) and from natural increase. There is a wide range of living conditions, together with social and spatial sorting.

Suburbanisation
People move from congested city centres to the margins of urban areas. Factories, retail and services also decentralise. Housing density in the suburbs is lower and there are more open spaces.

Counterurbanisation
The movement of people to dormitory settlements and commuter towns. Some move to rural areas. Governments may direct growth to new towns and areas designated for expansion. Counterurbanisation mainly includes movement to smaller urban settlements in rural districts.

Urban regeneration
Attempts by government and key stakeholders trigger the movement of people back into urban areas. Regeneration is achieved through housing renewal, improved services, redevelopment of old industrial sites, re-imaging through 'flagship' projects and the creation of new employment opportunities.

Check your understanding and progress at **www.hoddereducation.co.uk/myrevisionnotesdownloads**

Figure 9.1 Processes of urbanisation

Now test yourself

TESTED ◯

4   What is the difference between urbanisation and counterurbanisation?
5   Outline how governments can attract people back into urban areas.

Answers on p. 268

Making links

See Chapter 8 page 168 to link the processes in Figure 9.1 to changing characteristics of places.

Revision activity

Figure 9.1 outlines the characteristics of contemporary urban processes. For each process, add further notes on the:

a)   causes

b)   effects.

For example, for suburbanisation:

+   **Causes** include: the growth of public transport systems and increased use of private cars, both making commutes from the suburbs into the city centre easier.
+   **Effects** include: increased social segregation and funding being diverted from the inner cities to the suburbs.

The emergence of the megacity

There are four groups of large cities:

+   millionaire cities: population of 1 million plus
+   megacities: population of 10 million plus and perform services at national and international levels
+   world cities: large populations and have global influence in the service sector
+   metacities: conurbations with more than 20 million people.

World cities

World cities have major influences on the global economy and its business centres.

In addition, they are the locational headquarters of many TNCs and have a concentration of global financial service centres and producer services (law, advertising, accounting).

On the cultural side, world cities host major political, cultural and sporting events and have a reputation for world-class education.

World cities include New York, London, Tokyo, Sydney and Singapore.

Exam tip

Remember that Figure 9.2 represents an overview of some of the main processes; however, these processes will apply to rich and poor countries differently. Also, processes could be included in more than one column, for example globalisation.

Making links

See Chapter 8 page 166 for links to the concept of 'placelessness' and the homogenisation of cities.

Processes associated with urbanisation and urban growth

The processes leading to urban growth can be categorised into economic, social, technological, political and demographic. The flow chart in Figure 9.2 highlights some of the processes and lists additional factors leading to urban growth.

Exam tip

The processes leading to urbanisation will vary between countries – you must be prepared to apply your general understanding to specific contexts.

175

Factors leading to urban growth

Economic
- Cost of land
- Structural job changes
- New employment opportunities
- Industrialisation
- Economic development
- Affordable housing
- Opportunities for housing investment
- Potential for earning money in the informal sector
- Globalisation

Social
- Concentration in socio-cultural groups
- Geographical and social mobility
- Access to cultural and social participation and diversity
- Rural-to-urban migration (can also be driven by economic and political factors)

Technological
- More developed infrastructure
- Better connectivity
- Attraction of digital businesses

Political
- Regeneration schemes
- Re-imaging
- Planning decisions improving land use and making urban areas more attractive places to live

Demographic
- Population growth
- Attraction of urban areas to young, mobile populations

Figure 9.2 The processes leading to urban growth

'Push and pull factors' is a useful way of explaining the causes of rural-to-urban migration (see Table 9.1).

Table 9.1 Push and pull factors in rural-to-urban migration

Push factors	Pull factors
Population growth	Employment prospects
Pressure on the land for food and resources	Earning money in the informal sector
Environmental problems, e.g. desertification	Better quality of social provisions, e.g. health and education
High levels of disease	A perceived better quality of life
Decline in traditional farming practices	
Increase in cash crops	
Natural disasters, e.g. flooding, wars and civil unrest	

Revision activity

Create a diagram (it may have a similar layout to Figure 9.2) summarising the consequences of urban growth. Include the following and give a brief outline of each point:
+ Urban sprawl
+ A shortage of housing in cities in low-income countries
+ A shortage of affordable housing in cities in high-income countries
+ Unemployment and underemployment
+ Transport issues

Making links

See Chapter 8 page 170 for more on the re-imaging and rebranding of urban places.

Now test yourself   TESTED ◯

6 What is a a) megacity, b) world city, c) metacity?

7 How does globalisation lead to urban growth?

8 Explain the demographic processes leading to urban growth in low-income countries.

9 Define the 'informal sector' of an economy and explain why it is important.

Answers on p. 268

Check your understanding and progress at www.hoddereducation.co.uk/myrevisionnotesdownloads

Urban change

Deindustrialisation

In the UK in the second half of the twentieth century, people moved to cities for jobs in manufacturing, for example textiles (Manchester) and shipbuilding (Glasgow). A period of deindustrialisation followed when there were heavy losses in manufacturing jobs in British cities. Reasons included:

+ **mechanisation** – machines were introduced to complete the jobs of factory workers
+ **competition from abroad** – the rapidly industrialising countries of Taiwan, India and China had much cheaper labour, which reduced costs
+ **reduced demand** for certain products as new technologies were developed.

Figure 9.3 summarises the impacts of deindustrialisation in urban areas.

> **Exam tip**
>
> Remember to address the social and environmental consequences of urban decline, not just the economic impacts.

The impact of deindustrialisation on urban areas

Economic impacts	Social impacts	Environmental impacts
• Loss of jobs and personal disposable incomes	• Increase in unemployment	• Derelict land and buildings
• Closure of other businesses that support closing industry	• Higher levels of deprivation	• Long-term pollution of land from 'dirty' industries such as dye works and iron foundries remains a problem because there is no money for land remediation
• Loss of tax income to the local authority and potential decline in services	• Out-migration of population, usually those who are better qualified and more prosperous	• Deteriorating infrastructure
• Increase in demand for state benefits	• Higher levels of crime, family breakdown, alcohol and drug abuse, and other social problems	• Reduced maintenance of local housing caused by lower personal and local authority incomes
• Loss of income in the service sector as a result of falling spending power of the local population	• Loss of confidence and morale in local population	• Positive environmental impacts include a reduction in noise, land and water pollution and reduced traffic congestion
• Decline in property prices as out-migration occurs		
• Deindustrialisation leads to the **de-multiplier effect** in the urban areas affected		

Figure 9.3 The impacts of deindustrialisation on urban areas

Decentralisation

A process of decentralisation followed deindustrialisation.

+ Some people moved out of the congested core urban areas to lower-density housing developments on the fringe of cities.
+ New mechanised industry required more space for the automated production lines and parking for commuters, and it too found locations on the fringe of cities where there was more space and easier communications.
+ Services also decentralised:
  + Large retailers opened stores in new 'out-of-town' retail centres on the urban fringe with ample parking, for example the Metro Centre in the north east.
  + Offices decentralised to the same fringe locations as they followed their suburban workforces and took advantage of lower rents and less congestion.

> **Now test yourself**   TESTED ⬤
>
> 12 Differentiate between deindustrialisation and decentralisation.
> 13 How has decentralisation led to inequalities in cities in the UK?
>
> **Answers on p. 268**

> **Deindustrialisation**
> The loss of jobs in the manufacturing sector which happened in the UK in the second half of the twentieth century.

> **Revision activity**
>
> It is important to relate the impact of deindustrialisation to a named example to illustrate the points made in Figure 9.3.
>
> Using an example studied in class or from further research, make revision notes on how a named city has been impacted by deindustrialisation.

> **Now test yourself**
>
> 10 Outline one main cause of deindustrialisation.
> 11 Why has the impact of deindustrialisation been greatest in cities lacking economic diversification?
>
> **Answers on p. 268**
>
> TESTED ⬤

> **Decentralisation** The outward movement of people and activities from established centres.

The rise of the service economy

The decline of the manufacturing sector was accompanied by the rise of the service economy due to:

+ the growth of financial services to support new business developments and a more affluent population
+ more technologically advanced societies needing specialised services
+ leisure and retail services expanding with growing affluence and rising middle classes.

Urban policy and regeneration in Britain since 1979

Urban policy relates to the attempts by local and national governments to manage urban areas. Regeneration has been a focus of UK policy since the 1980s. Table 9.2 summarises the main regeneration policies since 1979.

Regeneration The revival of urban areas.

Table 9.2 A summary of urban policy in the UK since 1979

Urban policy	Details	Examples
1979–91 Emphasis given to property-led initiatives and the creation of an entrepreneurial culture	Greater emphasis placed on the role of the private sector to regenerate inner-city areas. Coalition boards were set up with people from the local business community. They were encouraged to spend money on buying land, building infrastructure and marketing to attract private investment.	Enterprise Zones Urban Development Corporations Urban Land Grants Derelict Land Grants City Action Teams
1991–97 Partnership schemes and competition-led policy	A greater focus on local leadership and partnerships between the private sector, local communities, the voluntary sector and the local authority. Strategies focused on tackling social, economic and environmental problems in run-down parts of cities, which now included peripheral estates.	City Challenge City Pride Single Regeneration Budget
1997–2000s Area-based initiatives	Many strategies in the 2000s focused upon narrowing the gap in key social and economic indicators between the most deprived neighbourhoods and the rest of the country. Local authorities were set targets to improve levels of health, education and employment opportunities, and funding was allocated to assist them in delivering government objectives.	Regional Development Agencies (RDAs) New Deal for Communities National Neighbourhood Renewal Strategy The Housing Market Renewal Programme
2011– Devolved responsibilities to English cities	Bespoke funding and agreements between central government and a city and/or Local Enterprise Partnerships allows the city to take responsibility for decisions affecting the city, e.g. relating to employment, economic growth, transport and health.	City Deals

Revision activity

From examples covered in class and/or by referring to Table 9.2, select and **evaluate** *two* contrasting regeneration policies (contrast in size, scope, focus, approach, i.e. 'top-down' economic regeneration or more recent 'bottom-up' approaches). Give a named location for each.

Organise your notes into a table outlining: the policy, advantages and disadvantages, and location/example.

Exam tip

It is important to be able to **evaluate** the impact of urban regeneration projects.

Check your understanding and progress at **www.hoddereducation.co.uk/myrevisionnotesdownloads**

Urban forms

REVISED

Urban form exists at all scales: street to regional. It refers to the physical characteristics that make up built-up areas, for example size, shape, density.

It is constantly evolving due to social, environmental, political, economic and technological developments.

Contemporary characteristics of megacities/ world cities

Megacities are a feature of modern urbanisation. Their characteristics are as follows:

+ They offer an expansive range of social services: health, welfare, education.
+ The environmental and planning impact of housing and infrastructure for dense populations is concentrated in one area.
+ They provide large and diverse employment opportunities.
+ They exist as centres for innovation.
+ They have better levels of education and healthcare which can improve the lives of the poor and empower women.

Making links

See page 176 for more on the emergence of megacities.

The term '**world city**' has been given to cities that have the greatest influence on a world scale. They include Tokyo, London, New York and Beijing. Their characteristics are summarised in Figure 9.4.

Figure 9.4 The characteristics of world cities

Now test yourself

TESTED

17 Describe the features of a world city.

Answer on p. 268

Urban characteristics in contrasting settings

The characteristic of size is not as important today: global influence is of greater importance.

Global influence can vary according to the spatial setting of the city, for example:

+ highly integrated cities that complement London and New York, such as Tokyo, Japan and Hong Kong
+ cities that link their region to the global economy, such as Santiago, Chile and Cairo, Egypt.

Physical and human factors in urban forms

Physical factors

+ **Physical factors** in urban forms relate to the initial reason for settlement location, such as relief and drainage.
+ Early industrial cities located near water for energy and transportation. Flat land was also important for ease of building and transport developments.
+ In poorer cities the steep land is often the marginal land where shanty towns develop.

Human factors

+ Following the initial location, **human factors** such as land prices tend to take over in terms of their effect on urban form.
+ The **bid-rent theory** seeks to explain how the location of urban land uses (retail, industry, residential) is determined by the willingness to pay high prices for central locations and reliance on accessibility (see Figure 9.5).

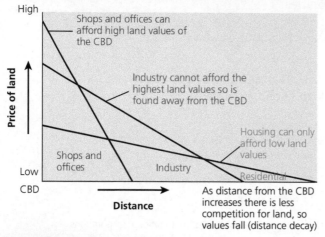

Figure 9.5 The bid-rent theory

Revision activity

Referring to OS maps, investigate the influence of physical factors (for example relief, drainage) and human factors (for example transport links, adjacent settlements) on a named urban area. Summarise your findings in a series of annotations to a base map or sketch map.

Spatial patterns of land use and the factors affecting them

The factors affecting spatial patterns of land use can be subdivided into categories, as outlined in Table 9.3.

Table 9.3 Factors affecting land use patterns in urban areas

Economic	Social	Political	Environmental
Land value – which activities can afford to locate in which parts of a city, e.g. retail, offices, warehousing, recreation	**Filtering** – as different social groups move between different areas of a city, social areas will change	**Impact of urban managers and their decisions** – planners, developers, government officials; these groups are particularly active in areas of regeneration (e.g. the work of bodies such as Urban Development Corporations in the 1980s)	**Relief** – flat land is easier to build on
Employment – job opportunities will lead to growth; unemployment decline	**Gentrification** – inward movement of more wealthy residents		**Drainage** – poor drainage leads to flooding; building on flood plains is an issue in the UK
Transport – ease of communication will be a locational factor for business, retail and workers	**Ethnicity** – ethnic groups may concentrate in certain parts of a city		**River sites** – have been attractive in the past as industrial locations
	Wealth – leads to spatial location of residents, it is related to cost and nature of housing		**Coastal locations** – offer cooler climate in hot countries; important for port activities and industrial growth

Now test yourself TESTED ○

18 What is bid-rent theory?

19 Outline how political factors affect land use patterns in urban areas.

Answers on p. 268

Check your understanding and progress at **www.hoddereducation.co.uk/myrevisionnotesdownloads**

New urban landscapes and the concept of the post-modern western city

Town centre mixed developments

The growth in online shopping has meant that many town centres have had to diversify in terms of their functions and this in turn has changed the landscape of many urban areas. There is now:

+ a wide range of leisure services, including neighbourhood cinemas, wine bars and restaurants
+ more communal open spaces such as squares or plazas
+ the promotion of street entertainment
+ flagship attractions such as at the Bristol Science Centre
+ a mix of apartments, conference centres and hotels to attract more urban tourists on city breaks.

Cultural and heritage quarters

+ Culturally led urban redevelopment projects were a feature of 1980s regeneration, for example Manchester's Northern Quarter.
+ These centres focus on the history of the area, as in Birmingham's jewellery quarter, which has a national reputation.

Fortress developments

+ These are up-market developments designed around security, protection and exclusion.
+ The inclusion of such developments in the housing stock of an urban area attracts affluent residents who will live, work and spend money in the city centre, leading to an upward multiplier effect.
+ Protection measures include CCTV, railings and fences, more street lighting and speed humps.

Gentrified areas

Gentrification is an essential part of housing improvement in urban areas and it has led to regeneration in many cities.

The process can be evaluated against a range of costs and benefits (see Figure 9.6).

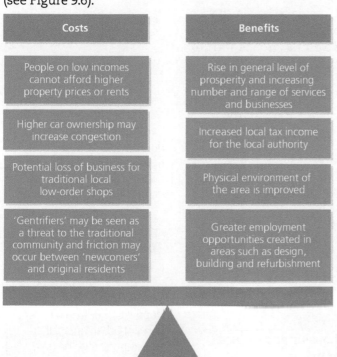

Costs	Benefits
People on low incomes cannot afford higher property prices or rents	Rise in general level of prosperity and increasing number and range of services and businesses
Higher car ownership may increase congestion	Increased local tax income for the local authority
Potential loss of business for traditional local low-order shops	Physical environment of the area is improved
'Gentrifiers' may be seen as a threat to the traditional community and friction may occur between 'newcomers' and original residents	Greater employment opportunities created in areas such as design, building and refurbishment

Figure 9.6 The costs and benefits of gentrification

> ### Exam tip
> Make sure you can relate each of the urban landscapes in this section to a named location. This could be linked to your chosen place study in Chapter 8.

> ### Making links
> You can draw links between these urban landscapes and 'insider' and 'outsider' perspectives on page 167 of Chapter 8.

> **Gentrification** The buying and renovating of properties in more run-down areas by wealthier individuals.

> ### Exam tip
> It is important to evaluate the impact of new urban landscapes. Figure 9.6 outlines the costs and benefits of gentrified areas. Edge cities also have costs in terms of potential social segregation and benefits in terms of retail development.

Edge cities

+ The result of urban sprawl, edge cities are a feature of urban landscapes in North America in particular.
+ They have good road communications and a range of services that have decentralised from the main urban centre.
+ In some countries they have been linked to social segregation as they are inhabited by more wealthy people, leaving the disadvantaged in the original city.

> **Edge cities** Self-contained settlements beyond the city boundary.

The concept of the post-modern western city

+ The post-modern term refers to changes in western society in the late twentieth century.
+ It is characterised by a mixed style of architecture with flagship structures such as The Shard in London.

> **Now test yourself** TESTED ⬤
>
> 20 What is gentrification?
> 21 Define the following: a) a fortress development, b) an edge city.
> **Answers on p. 268**

Social and economic issues associated with urbanisation

REVISED ⬤

Economic inequality

Economic inequality is a feature of urban areas. It refers to the wealth gap between rich and poor residents. Affluence can vary greatly across very short distances. Reasons include:

+ Housing types and values can vary within very small areas, leading to different social groups occupying the area.
+ Housing neighbourhoods change over time, for example large town houses may have been converted into flats for rent, and formerly run-down areas may have been gentrified.
+ When migrants arrive in a new country they are attracted to large urban areas for jobs, but they may find this difficult initially and start on low wages. They can afford only the cheapest housing so ethnic groups start to cluster in multicultural areas that persist for generations.

Multiple deprivation

This vicious circle of urban decay can lead to multiple deprivation for the residents.

Deprivation tends to be concentrated in:

+ large conurbations, particularly in inner cities
+ areas that previously had concentrations of heavy manufacturing
+ some coastal towns, for example Blackpool, Hastings
+ parts of east London.

> **Multiple deprivation** The lagging behind in a number of related aspects of life, such as employment, housing and services.

Measuring poverty and inequality

The **Index of Multiple Deprivation** is a UK government qualitative measure of deprivation for every neighbourhood in England. It is based on 37 separate socio-economic indicators, which are combined into seven indicators:

+ income
+ employment
+ health
+ education
+ crime
+ barriers to housing and services
+ living environment.

> **Exam tip**
>
> Make sure you are familiar with choropleth maps showing Index of Multiple Deprivation data in different parts of urban areas. It is also important to be able to account for the patterns shown.

Check your understanding and progress at **www.hoddereducation.co.uk/myrevisionnotesdownloads**

Social segregation

+ Inner-city areas have traditionally been the most deprived urban areas due to deindustrialisation and the resulting unemployment. Today, the pattern is more complex, with some high levels of urban poverty in peripheral estates.
+ Inequality and social segregation (the spatial concentration of the wealthy and the poor) can be more evident within a city than across a whole country.
+ Social segregation can also arise due to ethnicity.
+ Ethnic communities can become isolated from wider society as they maintain their own language and beliefs and have limited interaction with others outside their own community.
+ In American cities, the term 'ghetto' has been used to describe an area of a city dominated by an ethnic minority. Reasons for ethnic segregation are given in Figure 9.7.

Figure 9.7 Reasons for ethnic segregation in urban areas

Now test yourself   TESTED ◯

22 Define economic inequality.

23 What is the vicious circle of urban decay?

Answers on p. 268

Cultural diversity

Cultural diversity refers to the existence of a variety of cultural or ethnic groups within a society.

+ Urban areas are places where cultural diversity is a common feature.
+ **Globalisation** has increased the flows of people. **Diaspora** is a term used to describe a large group of people with a similar homeland who have settled elsewhere in the world, such as British people moving to Australia, referred to as expatriate communities.

+ Cultural diversity is a feature of urban areas because:
  + cities offer a wide range of employment opportunities
  + cities are often the first point of entry to a country for immigrants
  + there is often an established cultural diversity in cities.

Strategies to manage issues

A range of the strategies used to manage the issues outlined above is summarised in Figure 9.8.

Figure 9.8 Strategies used to manage the social and economic issues associated with urbanisation

Urban climate

REVISED

The impact of urban forms on local climate and weather

Urban areas have their own urban microclimate, which results in climatic differences between urban areas and the surrounding countryside in terms of:
+ temperature – an increase in both annual mean and winter minimum temperature
+ relative humidity – a decrease in annual mean, winter and summer levels
+ precipitation – an increase in quantity and days with >5 mm
+ visibility – increase in winter and summer fog
+ air quality – increase in dust particles
+ wind speed – decrease in annual wind speed and gusts.

Temperature and the urban heat island effect

Urban areas experience higher temperatures than surrounding rural areas – this is known as the urban heat island (UHI) effect.

This effect can be between 1°C and 3°C and will fluctuate depending on season, weather conditions, sun intensity and ground cover. Figure 9.9 summarises the UHI effect.

There are various reasons for this effect:
+ The building materials in urban areas absorb more heat than surfaces in rural areas and have a much lower albedo (the reflectivity of a surface).
+ Pollution outputs from industries and cars increase cloud cover, which absorbs outgoing radiation.
+ Water falling to the surface is disposed of quickly – this reduces evapotranspiration and means that there is more energy to heat the atmosphere.

+ Large concentrations of people, industries, homes and vehicles all release heat.

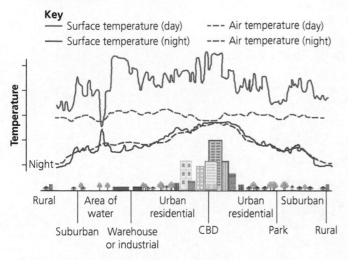

Figure 9.9 The urban heat island effect

Exam tip

Make sure you are familiar with the presentation of data on urban climate through a range of charts, graphs and maps, for example isotherm maps and line graphs relating change in air temperature to distance from the city.

Why does the urban heat island effect matter?
+ Human health – extremely high temperatures increase respiratory problems, heat stroke, pollen levels.
+ Disease – high temperatures increase the spread of vector and waterborne disease in poor cities.
+ Air conditioning systems use high amounts of energy – links to climate change.
+ There is an increased risk of the deterioration of parts of the urban fabric – historical monuments, more chemical weathering.
+ The impact of the urban heat island effect on climate change.

Factors leading to the intensity of the urban heat island effect
+ Growing urbanisation – concentrations of people, traffic and industrial activity.
+ Economic development (developing cities have large increases in building density and industrial pollution but low emission controls).
+ Anthropogenic heating (from vehicles, heating and air conditioning).
+ Air pollution, which increases particulate matter in the atmosphere, leading to heat retention.

Making links

See Chapter 1 for information on climate change and think about the links between climate change and urban climatic events.

Precipitation

Higher temperatures over urban areas lead to low pressure and an increase in rainfall. More intense convectional rainfall can also be common due to the heating of ground surfaces.

Figure 9.10 explains the cause-and-effect events leading to increased rainfall.

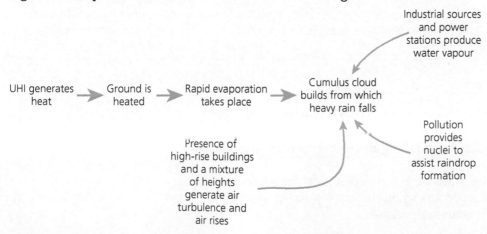

Figure 9.10 The causes of increased precipitation in urban areas

My Revision Notes: AQA A-level Geography Second Edition

Fogs and thunderstorms

In modern days this mainly applies to cities undergoing industrialisation, such as Beijing and New Delhi, where increased particles in the atmosphere act as condensation nuclei for fog formation. This is especially true in cities where there is no legislation (for example, Clean Air Acts). Figure 9.11 shows how thunderstorms develop in urban areas.

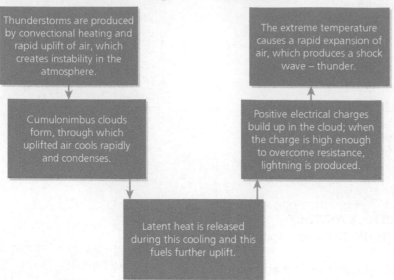

Figure 9.11 The formation of thunderstorms in urban areas

Wind

Urban structures have four effects on wind:

+ The varying heights and uneven surfaces of buildings produce frictional drag.
+ High-rise buildings channel air through the gaps or 'canyons' between them.
+ Upward convectional processes can draw in air from cooler surroundings.
+ When air flows between buildings the air movement is affected by the **Venturi effect**, where the pressure within the gap causes the wind to pick up speed, so urban areas may be subject to gusts of wind.

Now test yourself TESTED ◯

24 Define the following: a) urban heat island effect, b) the Venturi effect, c) albedo.
25 Outline one cause of the urban heat island effect.

Answers on p. 269

Revision activity

Draw a diagram to summarise the effects of urban buildings on winds. Include annotations that explain how a range of design features can impact urban winds and climate.

Exam tip

Remember that you can use simple, clear and well-annotated diagrams in the exam to explain a range of concepts and processes.

Air quality

Air quality in urban areas is often poorer than in rural areas as a result of:

+ particulate pollution
+ temperature inversions where cool sinking air can become trapped below a layer of warm air
+ photochemical smog, which is low-level ozone pollution associated with cars and pollution. This often occurs when there is a weather system of descending stable air, such as an anticyclone.

Table 9.4 summarises the main types of urban air pollution.

Check your understanding and progress at **www.hoddereducation.co.uk/myrevisionnotesdownloads**

Pollution-reduction policies

Strategies to reduce pollution are outlined in Figure 9.12.

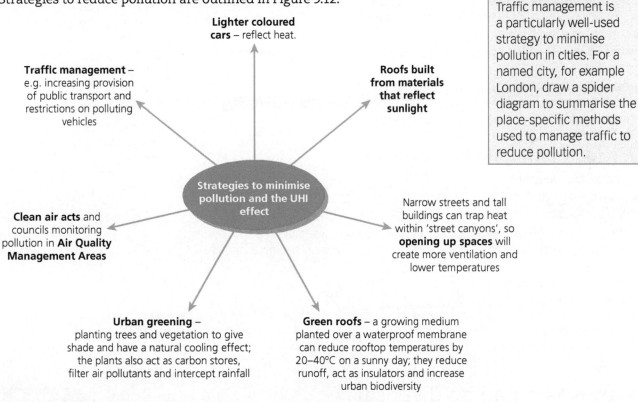

Figure 9.12 Strategies to reduce pollution and manage the UHI effect

Urban drainage

REVISED ◯

Urban precipitation

Precipitation falls in greater amounts in urban areas. The urban drainage system is designed to deal with surface runoff generated from impermeable surfaces as quickly and efficiently as possible through a system of underground pipes, sloping roofs, guttering and cambered roads.

My Revision Notes: AQA A-level Geography Second Edition

The impact of an urban catchment area on a flood hydrograph is summarised in Figure 9.13.

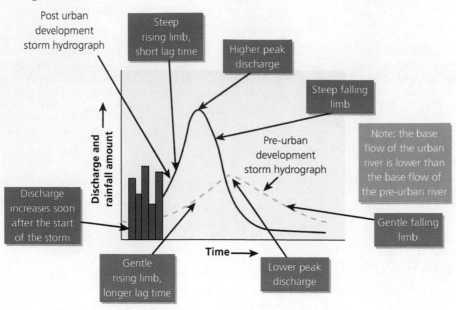

Figure 9.13 The flood hydrograph of an urban catchment area

Issues associated with catchment management in urban areas

Sustainable urban drainage systems (SUDS) are an attempt to manage surface water in urban areas by using natural processes in the landscape to reduce and control flooding. For example:
+ grass channels called swales, which direct water to storage areas in grass basins
+ permeable rock paving filters and stores water
+ water detention ponds and rain gardens reduce runoff and increase natural processes of storage and infiltration.

River restoration and conservation

See the revision activity.

Making links

See Chapter 1 page 17 for more on hydrographs and the impact of urban surfaces on the water cycle.

Revision activity

Use these terms to create a flow diagram explaining the effect of urban surfaces on the water cycle: runoff, infiltration, evapotranspiration, precipitation, waste water discharge, river discharge.

As part of each annotation you need to state whether the process is increased or reduced.

Revision activity

You need to be able to refer to a specific example of a river restoration and conservation project. Using an example studied in class, make revision notes under the following headings:
+ Reasons for the project
+ Aims of the project
+ Attitude and contribution of parties involved
+ Project activities
+ Evaluation of project outcomes

Urban waste and its disposal

REVISED

Sources of waste

+ As rates of urbanisation increase, the amount of waste produced in cities due to population increases.
+ Globally, the figure is about a 7 per cent increase year on year.
+ Waste generation varies significantly between cities.
+ In 2010, levels of waste were higher in cities in high-income countries.
+ Cities in low- and middle-income countries are set to see the largest increases in the future.

+ The amount of municipal solid waste is growing fastest in China.
+ A major concern is whether low-income countries will have the resources to deal with waste.

Sources of urban waste include:
+ household waste: food, plastics, paper, metals, ashes
+ industrial waste: packaging, construction and demolition materials, hazardous industrial waste
+ commercial waste: paper, plastics, food, glass, metals
+ construction and demolition: wood, steel, concrete, bricks, tiles.

Impacts of increasing waste generation include:
+ the costs of collecting and treating waste are high
+ waste is a large source of methane, a powerful greenhouse gas
+ untreated and uncollected waste can lead to health issues
+ many city authorities are struggling to collect increasing quantities of urban waste
+ some waste disposal may be unregulated, for example with no supervision by regulation laws.

Alternative approaches to waste disposal

Alternative approaches to waste disposal and their impacts are shown in Table 9.5.

Table 9.5 Approaches to waste disposal and their impacts

Approach to waste management	Description	Impact
Recycling	Materials are reprocessed into new products. This can save significant energy, e.g. a saving of 95% by the use of recycled materials to produce aluminium.	Large global market for recyclables Reduces the quantities of disposed waste Return of materials to the economy Can contribute to greenhouse gases
Trade	Global waste trade is the movement of waste between countries for treatment, disposal or recycling. The Basel Convention controls movement of hazardous waste.	Disposal of waste may not be controlled properly in countries with few guidelines and restrictions May create environmental issues in the recipient country
Incineration	General waste can be burned at high temperatures and under safe conditions.	Can reduce the volume of waste by up to 90% Can produce energy as an output Can lead to severe air pollution if not properly managed Quite expensive Air pollution and ash disposal are environmental concerns
Landfill (burial)	The burial of waste in man-made or natural excavations such as pits or landfill. In richer countries, the types of waste sent to landfill are strictly controlled.	Less regulated in poorer countries where it may just be a hole in the ground Gas produced can be collected to produce electricity Methane produced, also chemicals such as bleach and ammonia. Contamination of the atmosphere and groundwater is a major concern Landfill sites take up a lot of space and are unsightly

Other contemporary urban environmental issues

REVISED

Environmental problems in contrasting urban areas and management strategies

Urban environmental issues are a concern in the cities of both high- and low-income countries. However, it is in the poorer parts of the world that resources and legislation to deal with these issues may be lacking.

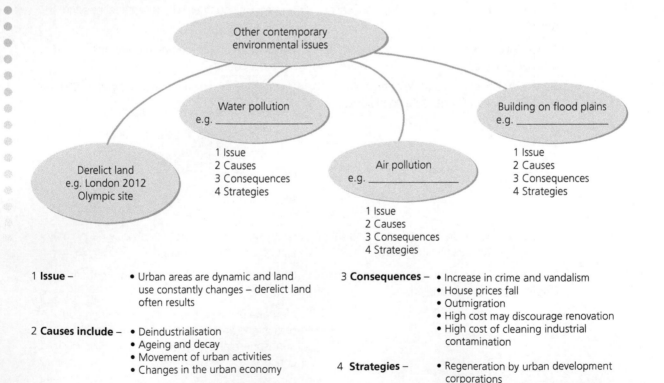

1 **Issue** –
- Urban areas are dynamic and land use constantly changes – derelict land often results

2 **Causes include** –
- Deindustrialisation
- Ageing and decay
- Movement of urban activities
- Changes in the urban economy

3 **Consequences** –
- Increase in crime and vandalism
- House prices fall
- Outmigration
- High cost may discourage renovation
- High cost of cleaning industrial contamination

4 **Strategies** –
- Regeneration by urban development corporations
- Focus on brownfield rather than greenfield sites
- Land remediation – removal of pollutants and contamination

Figure 9.14 Other contemporary urban environmental issues

Check your understanding and progress at **www.hoddereducation.co.uk/myrevisionnotesdownloads**

Sustainable urban development

Impacts of urban areas on local and global environments

Impacts of urban areas on local and global environments include:
+ consumption of resources
+ production of waste
+ pollution of air and water.

Ecological footprint of major urban areas

The ecological footprint is a measure of how much land it takes to provide the resources used by urban areas (farmland, forest, water, energy) and to dispose of the waste produced. It is measured in terms of the number of hectares of land needed to meet all the needs of one person.

> **Exam tip**
>
> Remember that for the higher-mark questions, you need to give details and facts from a specific geographical context, not descriptive evidence that could relate to any urban area.

The sustainable city

Dimensions of sustainability

The term 'sustainability' refers to improving quality of life while living within the Earth's carrying capacities. A sustainable city is a city that meets the following needs:
+ **health and welfare:** affordable housing, medical provision, protection from environmental hazards
+ **social needs:** access to education, equal opportunities and personal security
+ **economic needs:** secure jobs, access to a range of employment opportunities
+ **environmental needs:** fresh food and water, freedom from pollution
+ **political needs:** fair governance for all its people.

> **Exam tip**
>
> Sustainability features must focus on what is permanent and long term.

Features of sustainable cities

The concept of a sustainable city can be explained by reference to a systems model in which an unsustainable urban system with uncontrolled inputs and outputs is replaced by a more sustainable model where controlled inputs are recycled in order to reduce outputs (see Figure 9.15).

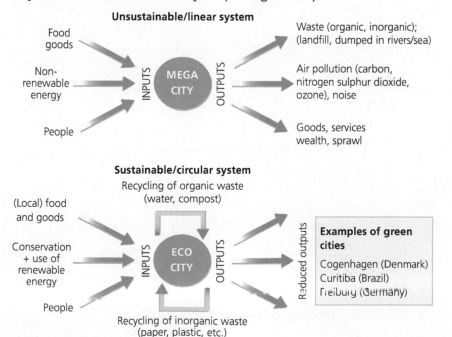

Figure 9.15 The city as a system

My Revision Notes: AQA A-level Geography Second Edition

Key features of sustainable cities include:
+ high usage of public transport
+ walking and cycling are safe
+ large areas of open space
+ use of renewable resources where possible
+ water is recycled wherever possible
+ affordable housing is accessible
+ energy-efficient building design
+ strong community links
+ cultural and social amenities are accessible to all.

The concept of liveability

This is a quality-of-life concept relating to the living conditions a city can provide for its residents. It means different things to different people depending on their priorities, for example career aspirations, political stability, a clean environment or cultural opportunities.

Opportunities and challenges in developing more sustainable cities

Opportunities and challenges vary according to level of development (see Figure 9.16).

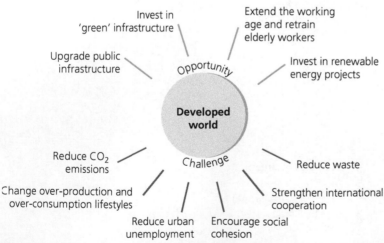

Figure 9.16 Opportunities and challenges in developing more sustainable cities

Check your understanding and progress at **www.hoddereducation.co.uk/myrevisionnotesdownloads**

Strategies for developing more sustainable cities

+ Investment in infrastructure, for example roads. Curitiba in Brazil has an integrated road transport system with many features to reduce the environmental impact of urban transport.
+ Waste management and reduction schemes, for example waste-for-food exchange in the favelas of Curitiba.
+ Provision of sustainable and affordable housing, such as the BEDZED development in the Greenwich Millennium Village, London.
+ Investment in renewable energy schemes, for example subsidies to residents for solar panels.
+ Adoption of a local currency.
+ Disaster risk reduction.
+ The active participation of different city stakeholders, e.g. government, residents and local businesses.

> **Now test yourself**   **TESTED** ◯
>
> 34 Define the concept of sustainability.
> 35 Explain how the concept of sustainability relates to urban areas.
> 36 How can the ecological footprint of cities be reduced?
> 37 What is liveability?
>
> **Answers on p. 269**

Case studies

Two contrasting urban areas, for example London and Mumbai

These case studies should:
+ illustrate and analyse the key themes in the contemporary urban environments unit, including:
  + patterns of economic and social well-being
  + the nature and impact of physical environmental conditions
+ make particular reference to:
  + the implications for environmental sustainability
  + the character of the study areas
  + attitudes of their populations.

What do I need to know?	Content and suggested revision methods

The character of the study areas, to include attitudes of the population	+ Summarise the key features of the **character** of the study area and include comments on the **attitudes of the population** in an annotated sketch map for each of your chosen cities.
	 + Include factors that add something more analytical than just a description. For example:
	   + London is a global financial centre that attracts direct foreign investment.
	   + Mumbai is now a megacity with the highest population density in the world – this places a strain on its urban infrastructure.
	 + Although it is useful to know background facts, such as population size and historical background, remember that restating knowledge will not earn you marks in the exam and will be a waste of both time and content.
	 + If you are using such facts, they must be analytical, for example:
	   + Population size is growing in Mumbai due to an influx of migrants.

What do I need to know?	Content and suggested revision methods
Patterns of social and economic well-being	Summarise bulleted notes in a table:

London	Mumbai
Economic well-being	**Economic well-being**
<ul><li>London has a third of the country's wealth but within that there are wide disparities.</li><li>In London, for every £1 of wealth owned by the bottom 10%, £172 is owned by the top 10%.</li></ul>	<ul><li>Poverty levels are high in Mumbai.</li><li>Many migrants enter the city seeking employment.</li></ul>
Social well-being	**Social well-being**
<ul><li>London's infant mortality rate has fallen by 40% over the last 10 years.</li></ul>	<ul><li>Around 60% of the population live in 'slums'.</li><li>In some areas of the Dharavi slum up to 1000 people share toilets, and water and electricity are not always available.</li></ul>

What do I need to know?	Content and suggested revision methods
The nature and impact of physical environmental conditions	Summarise facts in a flow diagram. This section requires you to **know** facts about environmental problems (i.e. the nature of environmental conditions) but you must relate them to their **impact.** For example:<ul><li>In Mumbai, nitrogen oxide pollution levels are dangerously high → this increases the occurrence of lung disease and other pollution-related illness and raises the death rate.</li><li>In London, there is an urban heat island effect → this impacts on health and also leads to a greater consumption of water and energy.</li></ul>
Strategies for environmental sustainability	Make revision notes on specific named strategies for environmental sustainability. Remember, **evaluation** of the named strategy is important, i.e.:<ul><li>How will it make the city more sustainable?</li><li>What disadvantages does it have?</li></ul>For example, in London, cycle superhighways and cycle hire schemes have been introduced since 2001 and there has been a 173% increase in cycling.<ul><li>This helps with sustainability as it reduces pollution and dependency on private and public transport.</li><li>However, there are concerns over cyclist safety.</li></ul>

Exam tip

The case studies for this chapter feature two contrasting urban areas. It would be useful for you to have an overview of those contrasts should an essay question require this.

Revision activity

Complete a simple table of similarities and differences of the two contrasting urban areas in the case studies.

Check your understanding and progress at **www.hoddereducation.co.uk/myrevisionnotesdownloads**

Exam practice

1 Analyse the causes of counter-urbanisation. [6]

2 Analyse the data shown in Figure 9.17. [6]

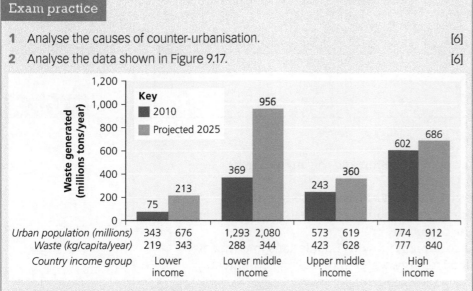

Urban population (millions)	343	676	1,293	2,080	573	619	774	912
Waste (kg/capita/year)	219	343	288	344	423	628	777	840
Country income group	Lower income		Lower middle income		Upper middle income		High income	

Figure 9.17 Urban waste generation by income level and year

Answers and quick quizzes online

3 With reference to named examples, assess the success of strategies for developing more sustainable cities. [9]

4 With reference to a river restoration and conservation project that you have studied, evaluate the project outcomes. [9]

5 With reference to examples, assess the extent to which economic factors have driven the emergence of megacities. [9]

Exam skills

Opportunities to practise geographical skills within this topic include:

+ use of statistical data and tests to interpret and manipulate data, for example:
  + census data and the Multiple Deprivation Index for socio-economic characteristics of urban areas
  + techniques such as frequency distributions, measurements of correlation
  + testing for the differences between data sets (Student's t-test and Mann-Whitney U test)

+ map analysis – to determine physical and human characteristics affecting the pattern of land use in urban areas
+ use of GIS data to compare the characteristics of different urban areas
+ fieldwork skills: data collection, recording, presentation and analysis.

Summary

+ Urbanisation is a significant global process, the progress of which can be understood through the comparison of world maps showing percentage urbanisation at different time periods.
+ A range of processes operates in urban areas, driven by economic, social and technological change.
+ Urban change is the result of a combination of deindustrialisation, decentralisation and the rise of the service economy.
+ Urban regeneration has been a focus of government policy in the UK since the 1980s.
+ Land use patterns of the 'urban form' exist at a variety of scales and for a range of reasons, which can be categorised into social, economic, environmental and political.
+ New urban landscapes have emerged in the twenty-first century: mixed developments in town

centres, cultural quarters, fortress developments, gentrified areas, edge cities and the concept of the post-modern western city.

+ A range of strategies is used to manage the social and economic issues associated with urbanisation.
+ Urban areas have an environmental impact, particularly on climate, drainage and waste disposal. Further environmental issues include air pollution, water pollution and dereliction.
+ Sustainability in the context of urban areas can be defined and measured. A range of opportunities and challenges exists to develop more sustainable cities.
+ Case studies of contrasting urban areas must be learned in detail and the nature of the contrast must be fully understood.

Introduction

REVISED

The environmental context

The link between population size and distribution and the environment centres on:

+ the **impact** of population growth on the environment through climate change, pollution, depletion of natural resources and the impact of human activities on ecosystems, ozone depletion
+ the **ability** of the environment to support growing populations in terms of basic resources, for example food, energy, water and land for settlements
+ **hazards** to population growth and development, for example the relationship between environmental variables and health – air quality and health, biologically transmitted diseases.

Key elements in the physical environment

Climate

The physical environment provides the most basic of human resources: food, energy, water.

Climate has a direct impact on food supply as it determines the global distribution of farming types and the quantities of food produced.

Adequate water supplies are vital to sustaining human populations and climate directly controls the quantity and distribution of rainfall.

Climate can also drive the level and nature of disease, for example malaria.

Soils

The key relevance of soils is their fertility as this impacts food production.

Densely populated areas often occur where there are fertile soils, for example the alluvial soils of the Nile Delta. Volcanic soils are also very fertile.

Modern technology allows the improvement of soils for food production through fertilisers and chemicals, but these have environmental consequences such as:

+ pollution
+ eutrophication
+ greenhouse gas emissions.

Resource distributions

Water

Water is needed for agriculture, industry and domestic use. Water supply is uneven due to variation in rainfall distribution, therefore some locations have a water surplus and others suffer from water scarcity. Even in areas where there is a reasonable amount of rainfall, high evapotranspiration rates mean that rainfall does not enter water stores.

Energy

The location of **energy** resources is determined by environmental factors, for example fossil fuels (gas in Russia, oil in the Middle East) – and the climatic conditions for renewable sources of energy (wind power, solar energy).

> **Making links**
>
> See Chapter 1 page 19 for more on the changes to the water and carbon cycles and the effect on food production.

Now test yourself   TESTED

1   Explain the link between population growth and:
    a) water pollution
    b) damage to wildlife and their habitats.
2   What is eutrophication?
3   What factors cause water scarcity in some areas?

Answers on p. 269

Key population parameters

The key parameters of the study of human populations are as follows:

+   **Population distribution:** the pattern of where people live can be applied to all scales – local to global. Global patterns reflect the location of physical resources, for example water and fertile land on the Nile delta.
+   **Population density:** the average number of people living in a specified area, usually expressed as people per km², for example Ganges valley, India (high population density).
+   **Total population number:** the raw figure of total population in an area at any scale. This can often be compared against population density as a very small population could be spread out over a large area and a large population could be concentrated in small areas.
+   **Percentage change in population:** change in population is the number of people added to or subtracted from a population in a year. The two components of change are **natural increase** and **net migration**. The change is expressed as a percentage figure from the beginning of the time period.

Exam tip

It is important that you're clear on the key measures used to study population change. Misinterpreting them could cost you marks in data response questions which require analysis of trends.

Key role of development processes

Population growth and development usually have negative impacts on the environment.

In general terms the fastest growth rates of population have been in the poorer parts of the world, but as countries develop and medical care improves, fertility rates fall.

In richer parts of the world population growth has been slow for several decades and some countries, such as Italy and Portugal, have even seen a small decline in population.

Global patterns of population numbers, densities and change rates

Maps of population data show the uneven nature of the spread of people across the Earth's surface. Total population numbers (raw data), **population density** and **population change** are useful measures. Figures 10.1 and 10.2 show population density in 2020 and, for comparison, 1994.

Key
(people/km²)

☐ No data ☐ 1–4 ☐ 5–24 ☐ 25–249 ■ 250–999 ■ 1,000+

Figure 10.1 World population density estimates, 2020

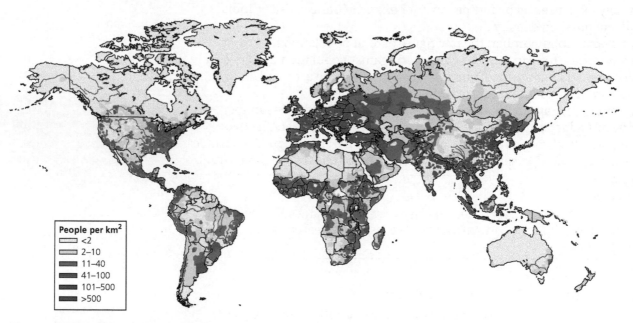

People per km²
☐ <2
☐ 2–10
☐ 11–40
■ 41–100
■ 101–500
■ >500

Figure 10.2 World population density, 1994

Now test yourself TESTED ○

4 Differentiate between population distribution and population density.

5 What are the components of a calculation of population change?

Answers on p. 269

Revision activity

1 Using Figure 10.1, list the factors that lead to areas of high and low population density. Remember that factors can be economic, environmental and social. You could add these as annotations on a copy of Figure 10.1.

2 Using Figures 10.1 and 10.2, list the main changes in the areas with high population density between 1994 and 2020. (Note the different scales on the keys.)

Check your understanding and progress at **www.hoddereducation.co.uk/myrevisionnotesdownloads**

Environment and population

Global and regional patterns of food production and consumption

Food production

+ Food production varies from place to place.
+ Climate (temperature, precipitation and growing season) and soils are the key environmental variables that determine food production.
+ Population size and skills, together with technology, capital and investment, are the key human variables for food production.
+ In 2015, food security had improved to a level where the world could produce enough food to provide every person with over 2940 calories per day.
+ However, food production and availability remain uneven.
+ In 2018, 820 million people still suffered with undernutrition.

Figure 10.3 shows the global pattern of food supply by calories available in 2013.

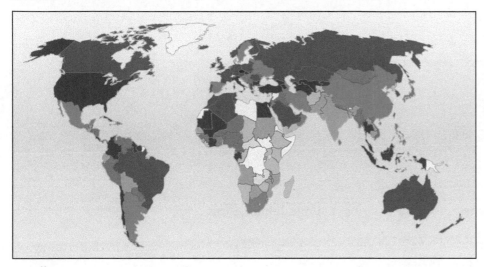

Key

| No data | 1,250 kcal | 1,500 kcal | 1,750 kcal | 2,000 kcal | 2,250 kcal | 2,500 kcal | 2,750 kcal | 3,000 kcal | 3,250 kcal | 3,500 kcal | 3,750 kcal | 4,000 kcal |

Figure 10.3 Global food supply by calories available, 2013

> **Food security** When all people at all times have access to sufficient, safe and nutritional food that will maintain a healthy lifestyle.
>
> **Undernutrition** Too little food to maintain a healthy body weight.

> **Exam tip**
>
> Remember good exam technique when *describing* patterns on maps – general points, specific high and low areas and identification of anomalies. Always quote supporting data. When asked to *analyse*, you must manipulate rather than repeat the data in order to add something.

> **Revision activity**
>
> Produce a bullet point summary of the global pattern of food production. Organise your notes into three sections: regions with high, sufficient and low levels of food supply. Be geographically specific, for example the continent of Africa: high levels in north Africa, for example Egypt; low levels in west Africa, for example Niger; and much of the continent with low levels or no data available.

Food consumption

Table 10.1 shows the global and regional per capita food consumption. It shows that:

+ the world has made significant progress in raising food consumption over the past 50 years
+ dietary energy has been steadily increasing on a worldwide basis
+ availability of calories per capita increased by almost 800 kcal per capita per day in developing countries between 1960 and 2015
+ large numbers are undernourished in some areas, for example Sub-Saharan Africa.

Rising incomes in industrialising countries have led to an increase in the consumption of meat and dairy produce.

Table 10.1 Global and regional per capita food consumption (kcal per capita per day)

Region	1964–66	2015	2030 (est)
World	2358	2940	3050
Developing countries	2054	2850	2980
Industrialised countries	2947	3440	3500
Transition countries	3222	3060	3180
Near East and North Africa	2290	3090	3170
Sub-Saharan Africa	2058	2360	2540
Latin America and the Caribbean	2393	2980	3140
East Asia	1957	3060	3190
South Asia	2017	2700	2900

Source: WHO

Now test yourself TESTED

6 If there is enough food to provide every person with over 2940 calories per day, why does undernutrition persist in some areas?

7 What human variables determine food production levels?

Answers on p. 269

Agricultural systems and agricultural productivity

Agricultural systems

Farms are open systems with:
+ inputs:
    + physical (for example climate, soil), cultural (for example, tenure, farm size)
    + economic (for example transport, technology)
    + behavioural (e.g. knowledge, experience)
+ farming processes (activity carried out to convert inputs to outputs)
+ outputs of animal products (for example beef, milk) and crops (for example cereal, vegetable, market gardening).

Losses can occur in the form of soil erosion, hazards and poor storage.

Agricultural systems can be put into categories, as shown in Figure 10.4.

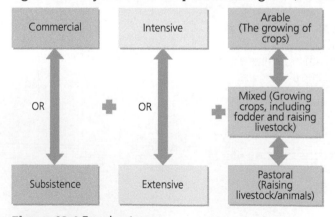

Figure 10.4 Farming types

> **Making links**
>
> See Chapter 7 for links to trade and the flow of agricultural products.

> **Revision activity**
>
> Figure 10.4 shows the main global farming types. Make sure that you are familiar with the meaning of commercial, subsistence, intensive and extensive farming by summarising a brief description as revision notes and listing a located example for each.

Agricultural productivity

Agricultural productivity represents the efficiency of the agricultural industry. It is measured in terms of a ratio of outputs to inputs, referred to as **total factor productivity** (TFP).

Check your understanding and progress at **www.hoddereducation.co.uk/myrevisionnotesdownloads**

Improving TFP is a crucial step towards producing more food more efficiently. This can be improved with:

+ disease-, drought- and flood-resistant crops
+ more efficient cultivation processes
+ better technology
+ high-quality animal feeds
+ favourable breeding practices for animals
+ expansion of land used for agriculture – extensification
+ additional inputs of machinery, fertilisers and pesticides – intensification
+ an increase in scientific research and development.

There can be widespread differences in TFP as it requires a high input of investment in technology and infrastructure.

Now test yourself   TESTED ◯

8   Define total factor productivity (TFP).
9   List four ways in which food productivity can be increased.

Answers on pp. 269–70

Relationship with key physical environmental variables – climate and soils

The two most important determinants of food production methods are climate and soils. Climate is a major factor affecting soil characteristics and, in turn, the natural vegetation of an area.

Figures 10.5 and 10.6 summarise the global pattern of world climatic and soil types.

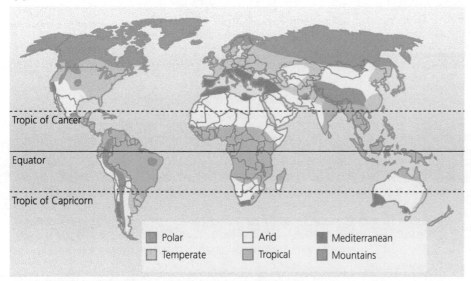

> **Exam tip**
>
> Make sure that you have practised how to compare distributions across different maps, for example world distributions or choropleth maps showing the spread of variables across the same area/ location.

Polar   Arid   Mediterranean
Temperate   Tropical   Mountains

Figure 10.5 Climate zones of the world

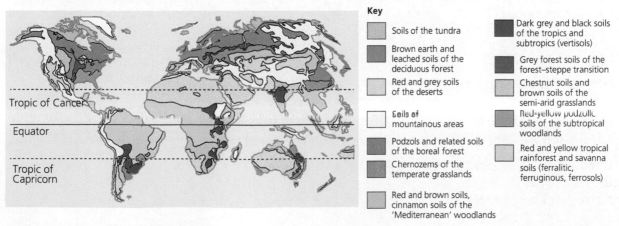

Key

Soils of the tundra

Brown earth and leached soils of the deciduous forest

Red and grey soils of the deserts

Soils of mountainous areas

Podzols and related soils of the boreal forest

Chernozems of the temperate grasslands

Red and brown soils, cinnamon soils of the 'Mediterranean' woodlands

Dark grey and black soils of the tropics and subtropics (vertisols)

Grey forest soils of the forest–steppe transition

Chestnut soils and brown soils of the semi-arid grasslands

Red-yellow podzolic soils of the subtropical woodlands

Red and yellow tropical rainforest and savanna soils (ferralitic, ferruginous, ferrosols)

Figure 10.6 Simplified world soil map

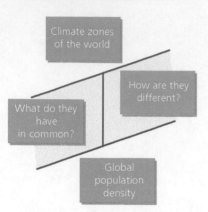

Figure 10.7 Revision diagram summarising comparisons

Characteristics and distribution of two climatic types

Tables 10.2 and 10.3 show two climatic types: polar regions and tropical monsoon.

Table 10.2 Climatic type 1: polar regions

Description: long, intensely cold winters, temperatures falling below −40°C, snow, strong winds. Land surface of glacial ice, snow or **permafrost**.	
Human activities	**Population numbers**
✦ Farming requires an input of technology – ground can be artificially thawed by a process of clearing vegetation, spreading manure, raising beds and insulating under polytunnels. ✦ Indigenous groups, for example Inuit, herd reindeer as a source of milk and meat.	✦ 13.1 million over 8 countries ✦ Population density of <4 per km² ✦ Most inhabit the tundra of North America and Eurasia ✦ Second half of twentieth century population grew due to improved health and discovery of natural resources ✦ Growth now slowing ✦ Population concentrated in larger settlements; indigenous people in scattered communities

Table 10.3 Climatic type 2: tropical monsoon

Description: typical of India and Bangladesh. Driven by a seasonal reversal of winds. Dry winds bring winter drought, average temperature 19°C. Southerly, summer winds bring hot and wet conditions – June to September, over 1500 mm rainfall and temperatures of 30°C.	
Human activities	**Population numbers**
✦ Rice – seedlings planted in flooded fields of the monsoon rains. Low mud walls retain the water. Mostly labour intensive. ✦ Strength of monsoon rain affects the yield and therefore the price and the economy of countries such as India.	✦ India – total population 1.32 billion, population density 446 per km², 17.8% of global population, 2016 ✦ Bangladesh – total population 162.9 million, population density 1252 per km², 2.19% of global population, 2016

Climate change as it affects agriculture

Sub-Saharan Africa region

People in the Sub-Saharan Africa region are likely to be the hardest hit by climate change due to their dependence on agriculture.

Over 60 per cent of the population live in rural areas working on small farms.

Potential impacts in Sub-Saharan Africa as a result of a 1.5°C global temperature rise include:
✦ long and frequent heatwaves
✦ drying soils
✦ less rainfall

+ frequent drought
+ reduced crop yields.

Asia-Pacific region

The Asia-Pacific region also has a high degree of vulnerability to climate change. There are 500 million rural poor in the region, many of whom are subsistence farmers.

Potential impacts of climate change in the Asia-Pacific region include:

+ warmer temperatures – predictions of a 0.5–2°C increase by 2030. This would lead to water stress in crops; yields of rice, maize and wheat in particular would decline
+ rising rainfall concentration and increase in summer rainfall – this would lead to flooding, which results in land degradation and soil loss. There is a belief that any increase in summer rain would not be enough to offset temperature increases
+ rising sea levels – this could be between 3 mm and 16 mm by 2030. This poses a particular risk of erosion and population displacement of coastal communities in low-lying areas, for example Bangladesh
+ more intense tropical cyclones – again risk of flooding and damage due to strong winds.

There are potential agricultural adaptations to be made in response to climate change:

+ heat- and stress-tolerant seed varieties
+ more efficient irrigation systems, for example drip irrigation
+ greater crop diversification to higher yielding varieties
+ mobile phone apps to connect to weather forecasting and agricultural advice services.

> **Land degradation** The deterioration of land.

> **Making links**
>
> See Chapter 1 page 29 for links to the causes of climate change.
>
> See Chapter 1 page 27 for links to the carbon cycle and impacts on soil properties.

> **Now test yourself**   TESTED ⭕
>
> 10 In what ways does food production in polar and tropical monsoon areas differ?
>
> 11 State three economic effects of climate change on agriculture.
>
> **Answers on p. 270**

> **Exam tip**
>
> When addressing issues such as climate change, remember to give a balanced response. In relation to agricultural productivity, some areas will have higher yields due to a warming climate.

Characteristics and distribution of two zonal soils

Healthy soils are essential to food production and food security.

A zonal soil is a major soil group often classified as covering a wide geographical region (see Figure 10.6). Tables 10.4 and 10.5 detail two zonal soils.

Table 10.4 Zonal soil 1: chernozem

Characteristics and distribution	Deep, black soils rich in organic content and minerals such as phosphorus, clay structure good for water retention, neutral to alkaline.
	Develop where there is a continental climate of cold winters and hot summers.
	Support tall, natural, grass vegetation.
Soil and human activity	The high fertility of these soils attracts modern agriculture such as arable cropping and cattle ranching.
	On the Steppes of Russia there is planting in both winter and spring.
	Crops on chernozem soils include wheat, barley, maize, soybean and potatoes.

203

Table 10.5 Zonal soil 2: red/yellow latosols of the tropical rainforest

Characteristics and distribution	Can be more than 40 m deep due to chemical weathering of parent rock in the hot, wet climate.
	High mineral content. Iron and aluminium compounds in surface layers give a rich, red colour.
	Not very fertile as the organic nutrients are stored in the vegetation.
	Decomposition is rapid in the warm climate and nutrients are quickly used by rapid plant growth.
Soil and human activity	Shifting cultivation is the main type of farming; burning vegetation releases nutrients to the soil.
	Land is then farmed for 2/3 years before being left to recover.
	This type of farming supports only very small populations.
	In recent years the rainforest has come under pressure from a variety of human activities, including cattle ranching and plantation agriculture.

Soil problems and their management

Soil erosion

Soil erosion is the wearing away of the top layer of soil. This is the most fertile soil as it contains organic matter and nutrients. Soil is eroded by both wind and water, as outlined in Figure 10.8.

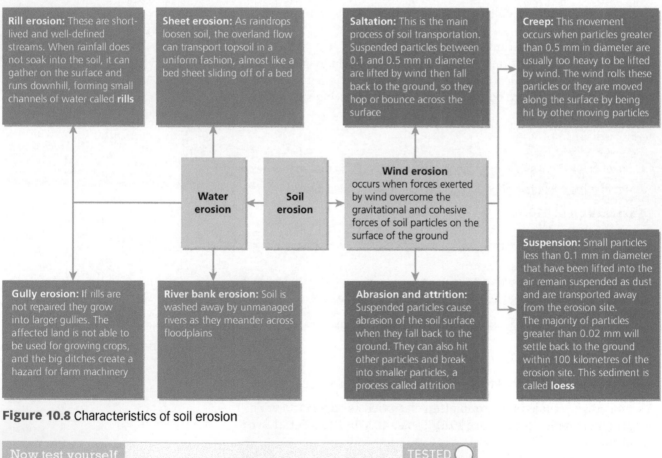

Rill erosion: These are short-lived and well-defined streams. When rainfall does not soak into the soil, it can gather on the surface and runs downhill, forming small channels of water called **rills**

Sheet erosion: As raindrops loosen soil, the overland flow can transport topsoil in a uniform fashion, almost like a bed sheet sliding off of a bed

Saltation: This is the main process of soil transportation. Suspended particles between 0.1 and 0.5 mm in diameter are lifted by wind then fall back to the ground, so they hop or bounce across the surface

Creep: This movement occurs when particles greater than 0.5 mm in diameter are usually too heavy to be lifted by wind. The wind rolls these particles or they are moved along the surface by being hit by other moving particles

Water erosion → **Soil erosion** → **Wind erosion** occurs when forces exerted by wind overcome the gravitational and cohesive forces of soil particles on the surface of the ground

Gully erosion: If rills are not repaired they grow into larger gullies. The affected land is not able to be used for growing crops, and the big ditches create a hazard for farm machinery

River bank erosion: Soil is washed away by unmanaged rivers as they meander across floodplains

Abrasion and attrition: Suspended particles cause abrasion of the soil surface when they fall back to the ground. They can also hit other particles and break into smaller particles, a process called attrition

Suspension: Small particles less than 0.1 mm in diameter that have been lifted into the air remain suspended as dust and are transported away from the erosion site. The majority of particles greater than 0.02 mm will settle back to the ground within 100 kilometres of the erosion site. This sediment is called **loess**

Figure 10.8 Characteristics of soil erosion

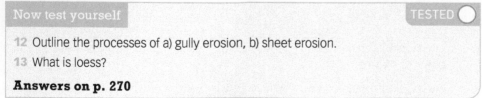

Now test yourself

TESTED ◯

12 Outline the processes of a) gully erosion, b) sheet erosion.

13 What is loess?

Answers on p. 270

Waterlogging

When water rather than air fills the pore spaces in soil it is said to be waterlogged. This impairs the ability of plants to respire and so their growth and development are reduced.

Making links

See Chapter 1 pages 13–15 for links to the drainage basin system and page 16 for the soil water budget.

Check your understanding and progress at **www.hoddereducation.co.uk/myrevisionnotesdownloads**

Waterlogging occurs due to:

+ an excess of water input (due to precipitation, irrigation water or flooding) and not enough percolation and/or evapotranspiration to remove the water from the surface levels
+ an excess of groundwater input, which is not matched by losses due to evapotranspiration.

Salinisation

Salinisation is a form of land degradation in arid and semi-arid climates. It is an increase in the amount of salts in the soil, which are brought to the surface when high rates of evaporation and transpiration combine with low precipitation and poor soil drainage.

Salinisation is a process that occurs naturally and is often made worse by human activity, for example the risk of salinisation is high where gravity flow methods of irrigation supply more water than crops can use.

Solutions to salinisation are described below:

+ Avoiding over-irrigation of crops by using techniques such as drip irrigation, soil moisture monitoring and accurate determination of water requirements.
+ Good crop selection – use deep-rooted plants to maximise water extraction.
+ Good soil management – maintain satisfactory fertility levels, pH and structure of soils to encourage growth of high-yielding crops.
+ Establishing and maintaining trees and shrubs on the land to maintain the water table.

Structural deterioration

Soil structures can be:

+ granular (typically sand, silt and clay in clumps through which water circulates)
+ blocky (large blocks of particles, which block the movement of water)
+ columnar (soil particles in vertical columns or cracks lead to poor drainage)
+ platy (where the particles are in thin horizontal sheets or plates; plates often overlap, leading to poor drainage).

Revision activity

Make a revision table to summarise a range of soil problems and potential management solutions as set out below.

Process	Resulting problems	Management solutions
Waterlogging – water rather than air fills the pore spaces in the soil.	Plant respiration is inhibited, and growth and development reduced.	Drainage pipes or ditches to prevent water being absorbed and held in the soil.

Exam tip

Practise a cause-and-effect sequence of explanation for soil problems – a leads to b, leads to c, etc.

Agriculture leads to changes in soil structure in two ways:

+ reduction in organic matter due to harvesting crops and rapid breakdown of organic matter in cultivated soils
+ ploughing leading to compaction. This impedes water percolation and can lead to gully erosion where farm vehicles form wheel ruts.

Now test yourself

TESTED ○

14 How does the soil type in a tropical rainforest affect farming?

15 Explain the process of salinisation.

Answers on p. 270

Strategies to ensure food security

A commonly used definition of food security comes from the FAO:

> Food security exists when all people, at all times, have physical and economic access to sufficient, safe and nutritious food that meets their dietary needs and food preferences for an active and healthy life.

From this definition there are four key components of food security, shown in Table 10.6.

Table 10.6 Dimensions of food security

Physical **availability** of food	This addresses the 'supply side' of food security and is determined by the level of food production, stock levels and net trade.
Economic and physical **access** to food	An adequate supply of food at the national and international levels does not, in itself, guarantee household-level security. Concerns about insufficient food access have resulted in a greater policy focus on incomes, expenditure, markets and prices in achieving food security.
Food **utilisation**	This is the way the body makes the most of various nutrients in food. Sufficient energy and nutrient intake is the result of good care and feeding practices, food preparation, diversity of diet and intra-household distribution of food.
Stability of the other three dimensions over time	Even if your food intake is adequate today, you are still considered to be food insecure if you have inadequate access to food on a periodic basis. Adverse weather, political instability or economic factors (unemployment or rising food prices) may have an impact on food security status.

Food security is a complex and contested term. Discussion points include the following:
+ The main issue is food distribution, not food production.
+ To what extent can future needs be met given projections in population growth and climate change?
+ What is the future role of trade in ensuring an equitable distribution of food?
+ The nutritional quality of food is a major issue across the development spectrum.

Exam tip

There are two main approaches to strategies to improve food security: increase supply and, on the demand side, diet change and potentially population control.

Exam tip

Be aware that some high-tech inputs to agriculture in poorer countries result in inequality and benefit only the wealthier farmers.

Strategies to ensure food security include:
+ short-term direct action to help the most vulnerable in terms of food security crises – food aid, appropriate technology, bottom-up projects, water aid projects
+ medium- and long-term building of resilience by:
  + improving agricultural productivity in poor countries
  + expanding rural infrastructure, for example to enable access to markets or gain better food storage
  + expanding knowledge of farming techniques and boosting research
  + reducing food waste
+ nutrition:
  + promoting healthy eating
  + educating and informing against overconsumption and poor diet choices.

Environment, health and well-being   REVISED

Global patterns of health, mortality and morbidity

Patterns of health are uneven. Figure 10.9 summarises the global pattern of mortality in an assumed group of 1,000 people. Mortality rates are higher in low- and middle-income countries and in the 70-plus age group.

Mortality The number of deaths in a population.

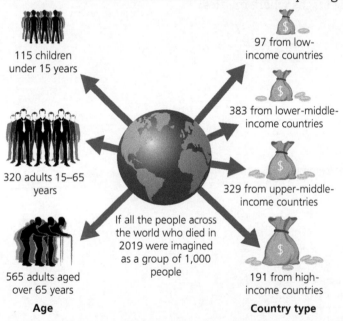

Figure 10.9 Distribution of mortality across age and country type, 2019

In terms of morbidity, non-communicable diseases were responsible for 71 per cent of deaths in 2018, a rise of over 60 per cent from 2000. There was an increase in the concentration of deaths from non-communicable diseases in low- and middle-income countries, which accounted for nearly 75 per cent of the deaths.

Morbidity The incidence of ill health.

Deaths from communicable diseases fell due to improvements in sanitation and health care.

In 2020–21, there is likely to be a sharp increase in the deaths caused by communicable diseases because of the Coronavirus pandemic.

Economic and social development and the epidemiological transition

As countries develop, the health and well-being of the nation should also improve as more money is available to spend on agriculture, health services and basic infrastructure.

Economic developments

+ Investment in agriculture to raise yields and farming efficiency in order to provide adequate good-quality food.
+ Improved infrastructure so that food can be stored and distributed efficiently and basic services such as energy, sewerage and clean water can reach the whole population.
+ Investment in the health service.

Social developments

+ Better education on sanitation, healthy diet and the spread of disease.
+ Advances in medical care and availability of basic medicines and vaccinations.
+ Better education and more opportunities to become fully trained health care professionals.
+ Reduced infant mortality rates.

Epidemiological transition

A model put forward by Abdel Omran in 1971 suggested that over time as a country develops there will be a transition from infectious diseases to chronic and degenerative diseases as the main cause of death.

The model, based on a line graph, is shown in Figure 10.10. It can be seen as a subsection of the demographic transition model in the stages where medical advances impact birth and death rates.

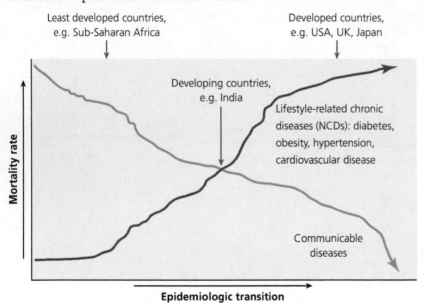

Figure 10.10 Line graph outlining epidemiological transition

Revision activity

Make a key terms list for this part of the topic. Include the following and any additional terms that you have difficulty with:
+ Disability-adjusted life years
+ Epidemiological transition
+ Morbidity
+ Mortality
+ Non-communicable disease
+ Well-being

Now test yourself

TESTED ◯

19 What is the difference between mortality and morbidity?

20 What is the epidemiological transition?

Answers on p. 270

Making links

The demographic transition model (DTM) on page 214 of this chapter provides a useful comparison to the epidemiological transition graph.

Check your understanding and progress at **www.hoddereducation.co.uk/myrevisionnotesdownloads**

The model was initially divided into three stages; a fourth was added in the 1980s. It has been suggested that the world's population is entering a fifth stage of transition, as shown with 5a and 5b in Table 10.9.

Table 10.9 The stages of epidemiological transition

Phase	Life expectancy	Change in socio-economic conditions	Causes of morbidity and mortality
1 Age of infection and famine	20–40	Poor sanitation and hygiene; unreliable food supply	Infections; nutritional deficiencies
2 Age of receding pandemics	30–50	Improved sanitation; better diet; advances in medicines/healthcare; reduced deaths from infectious diseases	Reduced number of infections; increases in occurrence of strokes and heart disease
3 Age of degenerative and man-made diseases	50–60	Increased ageing; lifestyles associated with poor diet, less activity and addictions	High blood pressure, obesity, diabetes, smoking-related cancers, strokes, heart disease and pulmonary vascular disease
4 Age of delayed degenerative diseases	c. 70+	Reduced risk behaviours in the population; health promotion and new treatments	Heart disease, strokes and cancers are main causes of mortality but treatment extends life Dementia and ageing diseases start to appear more
5a Age of inactivity and obesity	Potential reversal of life expectancy	Obesity increases at alarming rate; diabetes and hypertension increase; minority of the population meets recommendations for physical activity; ignore risk behaviours (especially in high-income countries)	Heart disease; stroke; congestive heart failure; diabetes; peripheral vascular disease
5b Emerging infectious diseases/ re-emergence of known diseases	Reversal of life expectancy; increase in mortality	Increased travel due to globalisation leads to increased transmission; increased population density; evolution of pathogens for which there is no known treatment or which are resistant to treatment; vaccine hesitancy	Pandemics – due to rapid evolution and spread of infectious diseases (e.g. SARS, COVID-19) Re-emergence of known diseases (e.g. TB, measles and whooping cough)

Due to variations in the pattern and pace of the transition, Omran identified three contexts to the model:

+ Classical/western model, for example western Europe, where a slow decline in death rate is followed by lower fertility.
+ Accelerated model, for example parts of Latin America, where falls in mortality are much more rapid.
+ Contemporary/delayed model, for example Sub-Saharan Africa where decreases in mortality are not accompanied by decline in fertility.

Revision activity

Annotate a copy of Figure 10.10 to show the change from communicable diseases to chronic degenerative diseases over time. Include:
+ some up-to-date country examples at various stages of the transition
+ a summary to reflect the evidence to support epidemiological transition.

Now test yourself   TESTED ◯

21 How can better education improve a country's health and well-being?
22 What is the basis for adding a fifth stage to the epidemiological transition model?

Answers on p. 270

Exam tip

It is the evaluation and application of models like the epidemiological model that is most useful. This is likely to be the nature of exam questions rather than just a description of their features. Make sure that you can offer some evaluation in terms of supporting evidence of the model and criticisms.

The relationship between the environment and incidence of disease

Climate

Extremes of climatic conditions can lead to the outbreak and spread of diseases.

Drought can lead to famine and illness related to lack of food, and floods can lead to the spread of water-borne diseases through poor sanitation.

A lack of sunlight and short days in the winter can lead to seasonal affective disorder (SAD). High pollen levels in the spring and summer can lead to hayfever and asthma.

The extremes of temperature in winter (pneumonia, flu) and summer (heat strokes) can lead to problems, especially for the very old, the very young and those with pre-existing illnesses.

Prolonged heavy rainfall can lead to the spread of vector-borne diseases.

Topography (drainage)

+ In countries or regions on flat floodplain land, there can be increased rates of diseases during periods of flood.
+ These include diarrhoea, dysentery and hepatitis A and E in countries such as India on the Ganges floodplain and China on the Yangtze.

Other environmental hazards

+ Earthquakes can have an impact on the incidence of disease, for example cholera and pneumonia outbreaks often occur in the cramped living conditions of emergency shelter in low-income countries.
+ There is a continued increase in the incidence of disease because of:
    + damage to health facilities and transport infrastructure
    + lack of health workers
    + increased exposure to disease vectors
    + lack of shelter
    + poor sanitation due to infrastructure damage.

Air quality and health

+ Air pollution is a health hazard in countries across the development continuum.
+ Polluted air causes cardiovascular and respiratory illness.
+ The WHO estimates that 80 per cent of air pollution deaths are related to heart disease and strokes, 14 per cent to respiratory infections and 6 per cent to lung cancer; 88 per cent of air pollution deaths are in developing countries, especially in Southeast Asia.
+ Fuelwood use in low-income countries for cooking and heating exposes households to respiratory infections, lung cancer and cardiovascular disease. Women and children are particularly at risk.
+ Most air pollution is beyond the control of individuals and must be addressed by decision makers at all scales, from governments to local authorities.

Water quality and health

Inadequate drinking water and poor water sanitation and hygiene cause 829,000 diarrhoeal disease deaths per year.

+ Parasitic worms in infested water lead to an estimated 240 million cases of schistosomiasis. Malaria has water-related vectors.
+ Human sewage is one of the main pollutants of water. Water Aid estimates that 800 million people live without safe drinking water. Sustainable Development Goal number 6 (post 2015) relates to safe drinking water and sanitation for all.

+ Environmental disasters such as oil leakage in the Nile Delta also contaminate water supplies.
+ In Lancashire in the summer of 2015, 300,000 people had to boil water for a month due to a water-borne bug (cryptosporidium), which causes sickness and diarrhoea, costing United Utilities £25 million in compensation.

Malaria is summarised in Figure 10.11.

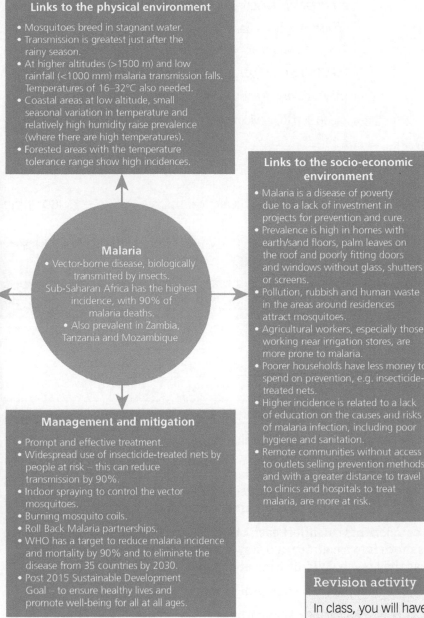

Links to the physical environment
- Mosquitoes breed in stagnant water.
- Transmission is greatest just after the rainy season.
- At higher altitudes (>1500 m) and low rainfall (<1000 mm) malaria transmission falls. Temperatures of 16–32°C also needed.
- Coastal areas at low altitude, small seasonal variation in temperature and relatively high humidity raise prevalence (where there are high temperatures).
- Forested areas with the temperature tolerance range show high incidences.

The impacts of malaria
Impacts on health
- Malaria kills a child somewhere in the world every minute.
- 90% of deaths are in Africa.
- The disease causes anaemia in children.
- Early stages – flu-like symptoms and high fever.
- Advanced stages – malaria may cause destruction of red blood cells, kidney failure, fluid on the lungs, convulsions, coma and ultimately death.
- Malaria impacts on school absences, decreased tourism and affects food production.

Impacts on economic well-being:
- Prolongs the vicious cycle of poverty.
- Economic costs to families to purchase medication and/or travel to health centres and hospitals for treatment.
- Costs to governments in drugs, education, medical staff.

Malaria
- Vector-borne disease, biologically transmitted by insects. Sub-Saharan Africa has the highest incidence, with 90% of malaria deaths.
- Also prevalent in Zambia, Tanzania and Mozambique

Links to the socio-economic environment
- Malaria is a disease of poverty due to a lack of investment in projects for prevention and cure.
- Prevalence is high in homes with earth/sand floors, palm leaves on the roof and poorly fitting doors and windows without glass, shutters or screens.
- Pollution, rubbish and human waste in the areas around residences attract mosquitoes.
- Agricultural workers, especially those working near irrigation stores, are more prone to malaria.
- Poorer households have less money to spend on prevention, e.g. insecticide-treated nets.
- Higher incidence is related to a lack of education on the causes and risks of malaria infection, including poor hygiene and sanitation.
- Remote communities without access to outlets selling prevention methods, and with a greater distance to travel to clinics and hospitals to treat malaria, are more at risk.

Management and mitigation
- Prompt and effective treatment.
- Widespread use of insecticide-treated nets by people at risk – this can reduce transmission by 90%.
- Indoor spraying to control the vector mosquitoes.
- Burning mosquito coils.
- Roll Back Malaria partnerships.
- WHO has a target to reduce malaria incidence and mortality by 90% and to eliminate the disease from 35 countries by 2030.
- Post 2015 Sustainable Development Goal – to ensure healthy lives and promote well-being for all at all ages.

Figure 10.11 Malaria: a biologically transmitted disease

Now test yourself   TESTED ○

23 Give three ways in which climate can impact the incidence of disease in the community.

24 How can domestic conditions lead to respiratory infections?

Answers on p. 270

Role of international agencies and NGOs in promoting health and combating disease

The World Health Organization, alongside various NGOs, is involved in promoting health internationally, as outlined in Table 10.10.

Revision activity

In class, you will have studied an example of a non-communicable disease and an example of a biologically transmitted disease. Using Figure 10.11 as a template, make revision notes on each of the examples studied in class.

Making links

See Chapter 7 page 157 for more on the agencies involved in global governance and their roles.

Table 10.10 Role of different organisations in promoting health and combating disease at a global scale

World Health Organization	Primary role to coordinate international health within the United Nations system.
	Specific early focus includes issues such as malaria, tuberculosis and nutrition, recently HIV, Ebola and Coronavirus.
	The WHO has responsibility for the classification, prevention and treatment of diseases, as well as coordinating responses to health crises.
	Criticisms include being overly bureaucratic and lacking practical front-line application.
Non-governmental organisations (NGOs)	Any non-profit-making association, working independently of government.
	They focus mainly on low-cost operations which involve local people.
	They are flexible and have freedom of response.
	In addition to providing health care and treatment, they are involved in training and research activities.

Revision activity

It is important to understand the role of different international agencies and NGOs in promoting health and combating disease.

Produce a revision summary in table form following the structure outlined below. Split the table into two parts – one for international organisations and one for NGOs.

Name of organisation	Nature of their role in promoting health and combating disease	Evaluation of contribution

Population change

REVISED

Factors in natural population change

There are two components of population change:
+ **Natural change:** difference between crude birth rate and crude death rate. Difference leads to natural increase or decrease.
+ **Migration change:** difference between immigrants (moving into an area) and emigrants (moving out of an area) leads to net migration change.

The two components can affect each other, as when a large number of young migrants move into a country and stay, then the natural increase will be affected as they reach child-bearing potential.

Key vital rates in population change are shown in Table 10.11.

Table 10.11 Population change: key vital rates

Birth rate	Live births per 1000 of the population per year. There is wide variation across countries, e.g. Niger 46, Japan 7 (2018).
Death rate	Deaths per 1000 of the population per year. Variation: Brazil 6, Russia 12 (2018). There is less variation across countries as medical advances spread more quickly than cultural change for birth rates.
Growth rate	From the natural change (difference between birth and death rate) the growth rate can be calculated by gaining a percentage value.
Total fertility rate	The average number of children that each woman of reproductive age will have. This is seen as a more accurate measure of future population change.
Net replacement rate	The number of children each woman needs to have to maintain the current population. In high-income countries it is around 2 and in low-income countries it is 3+.
Infant mortality rate	The number of children who die before reaching their first birthday. It has a wide use as it is age-specific, gives an indication of health care, gives an indication of wealth and has an impact on fertility rates.

Check your understanding and progress at **www.hoddereducation.co.uk/myrevisionnotesdownloads**

Cultural controls

Birth rate and fertility rate are affected by cultural factors. These include:
+ religion
+ gender preference for children (for example, some societies want males to work on farms)
+ status of women
+ marriage traditions.

Other socio-economic and political factors affecting birth and fertility rates include:
+ healthcare
+ education
+ population policies
+ political stability
+ economic conditions, for example growth rate, employment levels
+ affluence.

Models of natural population change and their application

Demographic transition model

A model represents an attempt to explain the processes of the real world.

The demographic transition model (DTM) is a graph that plots changes in birth and death rates over time and their impact on population growth. It was based on industrialised countries from the start of the industrial revolution. An updated version is shown in Figure 10.12.

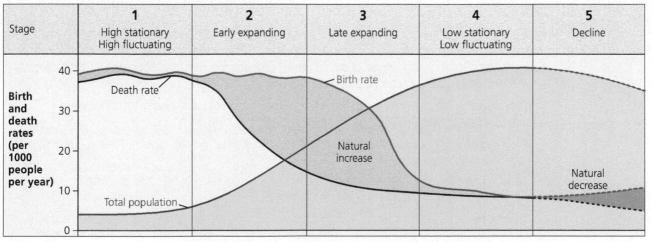

Figure 10.12 The demographic transition model

Application of the DTM to contrasting physical and human settings

The DTM can be applied to all countries as an attempt to explain the relationship between population change and development. Contrasts between different countries can be analysed to consider the extent to which human and physical factors affect demographic transition (see Table 10.12).

Table 10.12 Contrasting physical and human settings and demographic transition in Niger and Canada

Niger	Physical setting	One of the Sahelian countries with a very arid, sub-tropical desert climate.
		Much of the north and east of the country is Sahara Desert. The terrain is predominantly desert plains and sand dunes.
		The extreme south has a tropical climate near the edges of the Niger River basin.
		Non-desert areas in the south and west are threatened by periodic drought and desertification. Recurring drought events are a hazard and have occurred more frequently in recent years as a result of climate change.
	Human setting	It is mostly sparsely populated, especially in the north. There is a higher concentration of population in a band of towns and cities in the south, along the Niger River basin.
		The economy is very reliant on its primary sector, especially agriculture and mining. Most people live in rural areas and are subsistence farmers or nomadic pastoral herders.
		Niger is rich in uranium deposits and also started producing oil in 2010.
		More than 90% of the population are Muslim, coming from a number of different tribal groups.
		Droughts, failed crops, insect plagues and internal conflicts have led to food shortages, high food prices and hunger for many.
	Application to the DTM	High birth rates (largely as a result of religious and cultural beliefs) and relatively low and falling death rates put Niger in stage 2 of the model.
		Death rates are much lower than would be expected for stage 2. This is partly due to the country's young population structure and also because the government has made great strides to improve child mortality rates by reducing hunger and malnutrition and improving health care. It has been supported by NGOs such as Save the Children and the Eden Foundation.
Canada	Physical setting	Comprises a wide range of climatic types including Arctic, temperate continental and temperate maritime.
		The west of Canada is mountainous and the southern central area features rolling fertile plains.
		The north is a rugged and mountainous wilderness of taiga and tundra. Canada is extremely rich in mineral resources.
	Human setting	It is a very large country with a relatively small population of around 35 million people, so on a national scale it is sparsely populated but highly urbanised, with a concentration of population in the large cities in the southeast of the country, bordering the USA. Smaller concentrations appear in the southern parts of central provinces and on the west coast.
		The economy is based very much on its tertiary sector, particularly financial services and manufacturing, though it also has thriving mining and oil industries.
		It is a multicultural society, mostly welcoming to immigrants and tolerant of different cultures, languages and traditions.
	Application to the DTM	Canada has a low birth rate and death rate with a low natural increase. This places its demographics fairly clearly in stage 4 of the model.
		Despite being a wealthy country, Canada has not progressed into stage 5. Indeed, the natural increase is getting slightly larger. The main reason is that Canada encourages some controlled immigration. Having plenty of space and being rich in resources, it can cope with increased population and wants to keep a balanced structure that avoids becoming an ageing population, as in other equally rich countries.

> **Exam tip**
>
> Pure description of models will not earn you marks; it is important to be able to critically evaluate a model's application and usefulness in understanding current patterns of population change.

214

Check your understanding and progress at **www.hoddereducation.co.uk/myrevisionnotesdownloads**

Age–sex composition

Population structure refers to the age distribution and sex composition of populations. It can be applied to all scales from local to regional to national. Mostly it is used at the national scale.

The structure is represented in a graph known as a **population pyramid**, annotated to show how it is interpreted in Figure 10.13.

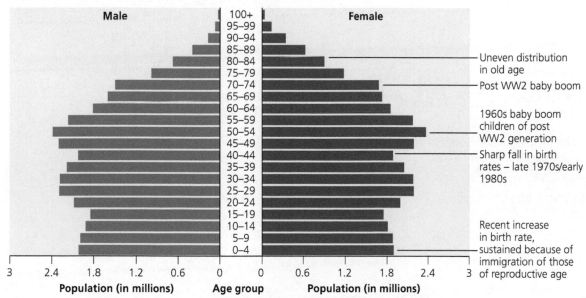

Figure 10.13 Population pyramid showing the population structure of the UK, 2018

Population pyramids:

+ show the effects of large-scale migration
+ show past changes in population
+ can be used to predict short-term and long-term change in population
+ show the effects of war, disease and famine
+ indicate life expectancy for different genders
+ can be related to stages in the DTM, as in Figure 10.14.

215

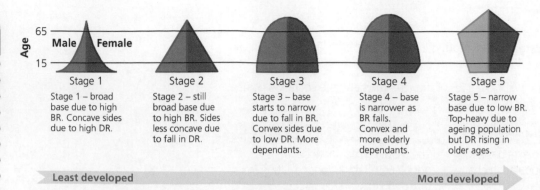

Figure 10.14 Population pyramid sketches for each stage of the DTM

Revision activity

Match each of the population pyramids in Figure 10.14 to a named country and annotate its features on a copy or sketch of the pyramid.

Exam tip

The population pyramid is a snapshot in time. You may need to predict changes and implications, for example as a high birth rate (0–4 range) enters the reproductive age.

Implications of different population structures for resources and development

There are problems and benefits in both young and older populations, as summed up in Figure 10.15.

Problems

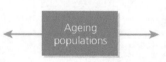

- Increased demand of maternal and early years healthcare.
- Increased demand for schooling.
- Long-term investment needed to resource a growing population as young reach reproductive age.

Youthful populations

Benefits

- In the long term, more human resources.
- Favourable political and economic conditions can foster growth.
- Growing market for goods internally as the population expands.

- High cost of elderly healthcare.
- Pensions cost more – often the costs are borne by the smaller, economically active population.
- Tax on workers may rise.

Ageing populations

- Some target markets increase, e.g. residential care, leisure.
- Fewer people of working age so lower unemployment.
- Increase in elderly contributions, e.g. looking after grandchildren and volunteer work in the local community.

Figure 10.15 The problems and benefits of youthful and ageing populations

Now test yourself   TESTED ◯

28 Sketch the population pyramids and characteristics of a youthful and an ageing population.

29 What are the economic implications of a youthful and an ageing population?

Answers on pp. 270–71

Exam tip

Remember that government attitudes are an important component of a country's response to a youthful or an ageing population. Keep up to date on this as government policy changes, for example amendments to China's one-child policy.

Concept of dependency ratios

The dependency ratio is a measure of the dependency of the non-working (0–14 and 65+) on the working, economically active (15–64) population. It is expressed as a simple equation:

$$\frac{\text{young dependants (0–14)} + \text{elderly dependants (65 and over)}}{\text{working population (15–64)}} \times 100$$

Check your understanding and progress at **www.hoddereducation.co.uk/myrevisionnotesdownloads**

For the UK the figure is 58.62, meaning that for every 100 working people there are around 59 people dependent on their earnings.

Concept of the demographic dividend

This is the benefit a country gets when the working population outgrows its dependants.

The low dependency for the elderly and the very young leads to an economic boost.

This has occurred in economies of the Asian Tigers where investment in education and employment has made the most of the demographic dividend.

The occurrence of natural resources, such as in Brazil and China, will also benefit the country. Political stability and scope for change are also required.

International migration

The geographical definition of migration is 'the movement of people across a specified boundary in order to establish a new, permanent or semi-permanent residence'.

Migration impacts population change for both the country of **origin** and the **destination** country.

There are different categories of migration: **labour migration, family migration, humanitarian migration**.

In 2019, the UN published the following findings:
✦ 66 per cent of all international migrants lived in 20 countries.
✦ 33 per cent of all international migrants came from just 10 countries, including India, Mexico and China.
✦ 74 per cent of all migrants were 20–64 years old.
✦ Women and girls made up approximately 48 per cent of all international migrants.

Types of migrants

Types of migrants are outlined in Figure 10.16.

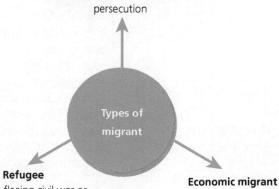

Asylum seeker
A person who flees their country of origin and applies for asylum on the grounds that they have a well-founded fear of death or persecution

Types of migrant

Refugee
A person fleeing civil war or natural disasters but not necessarily persecution; legally, a refugee is an asylum seeker with a successful asylum claim

Economic migrant
A person seeking employment in another country

Figure 10.16 Types of migrant

Revision activity

Referring to the Asian Tiger economies, produce a flow diagram to explain the causes and consequences of the demographic dividend. The flow diagram has been started below:

Previous high birth rate numbers move into the economically active age group → There are fewer elderly dependants due to previously low life expectancy →

Now test yourself

30 Define the demographic dividend.

31 Why do countries need to invest in order to take advantage of the demographic dividend?

Answers on p. 271

TESTED

Environmental and socio-economic causes and processes

People migrate due to a range of **push** (reasons for leaving) and **pull** (attractions of the destination) factors, as shown in Figures 10.17 and 10.18. Migration may also be forced (when there is no option) or voluntary.

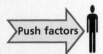 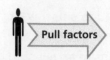

ORIGIN	DESTINATION
'Forcing' factors	**Associated with voluntary migration**
• War, conflict, political instability	• Better quality of life, standard of living
• Ethnic and religious persecution	• Varied employment opportunities, higher wages
• Natural and man-made disasters such as earthquakes, tsunamis, drought, famines	• Better healthcare and access to education services
	• Political stability, more freedom
	• Better life prospects
Socio-economic conditions	**For retirees:**
• Unemployment, low wages or poor working conditions	• Specific type of environment with a range of services to cater for their needs
• Shortage of food	

Figure 10.17 Push and pull factors

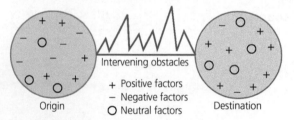

- For people to move, they need to be 'pushed' from their country of origin and 'pulled' to another country.
- The majority of migrants are voluntary movers, doing so for largely economic reasons.
- The negative factors at the origin are the 'push' factors; the positive ones at the destination are 'pull' factors.
- Migrants have to evaluate these factors and obstacles before they move.
- Intervening obstacles might include: travel costs, family pressures, language barriers, misinformation, immigration controls, bureaucracy, border controls.

Figure 10.18 Lee's push-and-pull model of migration

Implications of migration

Table 10.13 outlines the various implications of migration.

Table 10.13 Implications of migration

Implications at origin (home country)	Implications at destination (host country)
Demographic implications	
Lower birth rates; people of childbearing age leave	Balances population structure, if previously ageing population
Population structure – ageing population remain; population unbalanced	Migrants in reproductive age groups means increase in birth rates
Loss of male population of working age	Increase in male population of working age
Social implications	
Advantages	**Advantages**
Reduced pressure on health care (see health implications)	Cultural advantages of new foods, music, fashion, etc.
Reduced pressure on education	**Disadvantages**
Disadvantages	Pressure on maternal and infant health care
Loss of traditional culture	Pressure on schools (particularly primary)
Break-up of family units	Young male migrants may create social problems
Break-up of communities	Can give rise to ethnic and racial tensions
May lose qualified workers such as doctors, nurses and teachers	Possible increase in crime due to poverty
	Segregation of migrants into certain areas
Economic implications	
Advantages	**Advantages**
Reduced pressure on food, energy, water, etc.	Overcomes any labour/specific skill shortages
Less unemployment	May provide cheap labour who work longer hours
Remittances sent back home by migrants	Working migrants spend money/pay taxes
Migrants develop new skills, which they can bring back home	Increases size of workforce – can provide economic boom and multiplier effect
Disadvantages	Reduced dependency – 'demographic dividend'
Lose better educated/most skilled from workforce	**Disadvantages**
Creates dependency on remittances	Pressure on jobs/unemployment
Less agricultural and industrial production	Resentment towards migrants in time of recession
Decline in services – not enough people to support them	
Political implications	
Pressure to redevelop areas in decline	Pressures to control immigration
May introduce pro-natal policies	Rise of anti-immigration political parties
	Growth of right-wing, racist organisations
Environmental implications	
Farmland, buildings and sometimes whole villages may be abandoned	Pressure on land for development – roads, housing, other infrastructure
Less environmental management	Increased demand for energy, water and food puts pressure on natural resources
Health implications	
Migrants leave areas where infectious diseases are endemic or sometimes epidemic	Increase in infectious diseases transmitted by/to migrants from areas of different disease prevalence
Less pressure on limited health services but …	Increased pressure on health services because of rise in infectious diseases
… demographics of migration mean that the most vulnerable (children, elderly and poor) remain at risk	Increased pressure on health services to treat non-communicable/chronic diseases – the notion of 'health tourism'

Principles of population ecology and their application to human populations

Population growth dynamics

+ In ecology, population is determined by birth rates and death rates.
+ **Biotic potential** is the natural reproductive potential of a species, which controls birth rate.
+ Environmental limiting factors, which are often density dependent, control death rate.

Global population is over 7 billion. This exponential growth of human population over the past 200 years has been attributed to improvements in medical care, food supply and sanitation. In this way the human population has overcome much of the environmental resistance to its growth.

The relationship between human population and resources can be expressed through three key concepts:

+ **Overpopulation:** too many people for the available resources; continued increase in population reduces the standard of living. It is characterised by unemployment, under employment, outward migration, and insufficient food and resources.
+ **Underpopulation:** too few people to use resources effectively for a given level of technology. It is characterised by low unemployment, in-migration and good living conditions.
+ **Optimum population:** a population size where the best standard of living can be achieved given the resources in the area.

These concepts can be applied to all scales. A simple graph summarises the relationship (see Figure 10.19).

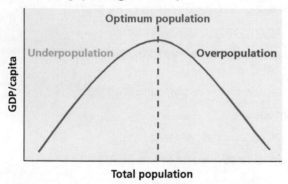

Figure 10.19 Relationship between population size and standard of living

Carrying capacity and ecological footprint

Carrying capacity is a term used in ecology. It refers to the size of population a given environment can support.

In human terms it relates to levels of consumption since a set amount of resources can carry a larger population with low levels of consumption but a smaller population with higher levels of consumption.

The calculation of **total productive biocapacity** gives an idea of the Earth's carrying capacity. The result is often expressed as an **ecological footprint** – the global hectares available for each person on the planet.

For every society there will be a theoretical optimum population in terms of its size and structure. It can be adapted in light of changes in resource supply for each person on the planet.

> **Exam tip**
>
> Concepts of over-, under- and optimum population and ecological footprints can be applied at any scale.

> **Now test yourself**
>
> 33 Explain the term 'optimum population'.
>
> 34 Differentiate between the terms 'carrying capacity' and 'ecological footprint'.
>
> **Answers on p. 271**
>
> TESTED ⬤

Check your understanding and progress at **www.hoddereducation.co.uk/myrevisionnotesdownloads**

Implications of carrying capacity and ecological footprint

The negative environmental implications for growing ecological footprints include:

+ exacerbation of global warming
+ more land taken for settlement, industry and transport
+ degradation of natural ecosystems
+ increased threat of species' extinction
+ over-cultivation and over-grazing reducing land and soil quality
+ depletion of fish stocks beyond recovery
+ depletion of fresh water supplies.

Population, resources and pollution model

The **population, resources and pollution model** (PRP), shown in Figure 10.20, provides a fundamental understanding of the relationship between humans and their environment.

The model adopts a systems approach to explain that the acquisition of resources alters ecosystems. The resources are used and the conversion into energy or finished products results in pollution.

+ Positive feedback amplifies changes, making the system less balanced.
+ Negative feedback counters any change, holding the system in a more stable state.

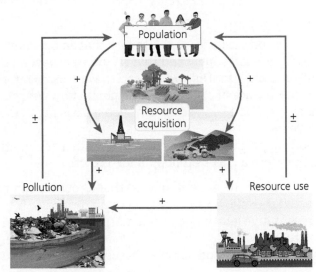

Key:
+ signs indicate a 'positive feedback' loop in which one activity increases another
− signs indicate a 'negative feedback' loop in which one activity reduces another

Figure 10.20 The population, resources and pollution model

Figure 10.21 summarises some of the challenges resulting from population growth.

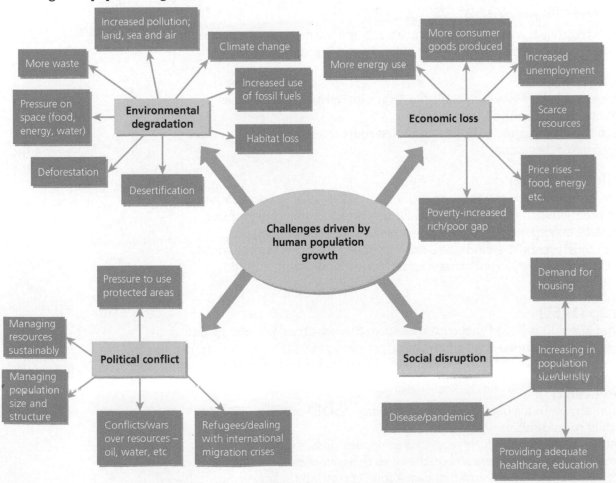

Figure 10.21 Challenges faced by the modern world driven by population growth

My Revision Notes: AQA A-level Geography Second Edition

35 Outline one negative feedback in the PRP model.
36 Outline the link between each of the following and population growth:
   a) desertification, b) pandemics, c) increased affluence.

Answers on p. 271

Contrasting perspectives on population growth

Malthus (1798)

Malthus based his ideas on the theory that an optimum population exists in relation to food supply and that an increase beyond this will lead to 'war, famine and disease'. His two principles were as follows:

+ In the absence of checks, human population will grow at a geometric rate: 1, 2, 4, 8, 16 and so on. On such a basis population will double every 25 years.
+ Food supply at best can only increase at an arithmetic rate: 1, 2, 3, 4, … and is therefore a check on population growth.

Given a limit to the amount of food that a country can produce, Malthus suggested the following checks to population growth:

+ preventative (abstinence from marriage, delay in the time of marriage and pregnancy)
+ positive (lack of food, famine, disease and war) checks to population growth.

Since Malthus's theory was put forward, food production has increased through:

+ HYV (high-yield variety) crops
+ new foods such as soya
+ use of agrochemicals
+ use of greenhouses and polytunnels
+ land acquisition, for example drainage of wetlands.

Malthus's theories have gained support recently among demographers known as neo-Malthusians based on evidence of famines, wars and water security.

Ehrlich (1968)

+ Ehrlich, in his controversial book *The Population Bomb*, predicted mass starvation of millions by the 1970s.
+ When this didn't come true, he stated that more intensive agriculture had helped avoid disaster.

The Club of Rome

+ Their 'Limits of Growth Model' published in 1972 is based on population, food production, industrialisation, pollution and resource consumption.
+ It focuses on exponential population growth and its consequences.
+ After adjustments, the model still predicts that population growth and resource consumption will worsen conditions for humanity within 40–50 years.

Boserup (1965)

+ Esther Boserup believed that countries have the resources, knowledge and technology to increase food supply in response to growth in population and that population growth is needed to trigger such advancements.

Simon (1981)

+ Simon argues in his book *The Ultimate Resource* that the only resource the Earth is running short of is people.
+ He states that resources have increased as population has grown, that is raw materials have become less scarce, the air in rich countries is safer to breathe, water cleanliness has improved, the condition of crop land is improving, and food production increase will keep pace with population growth.

Check your understanding and progress at **www.hoddereducation.co.uk/myrevisionnotesdownloads**

Global population futures

REVISED

Health impacts of global population change

Figure 10.22 summarises some of the health impacts of the change in the global population.

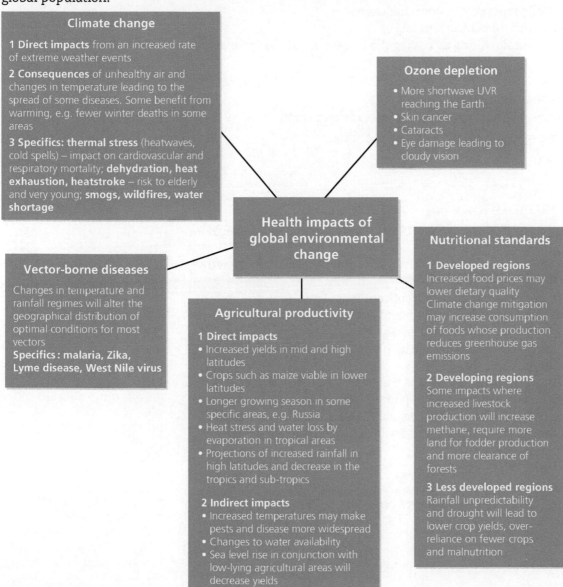

Climate change

1 Direct impacts from an increased rate of extreme weather events

2 Consequences of unhealthy air and changes in temperature leading to the spread of some diseases. Some benefit from warming, e.g. fewer winter deaths in some areas

3 Specifics: thermal stress (heatwaves, cold spells) – impact on cardiovascular and respiratory mortality; **dehydration, heat exhaustion, heatstroke** – risk to elderly and very young; **smogs, wildfires, water shortage**

Ozone depletion

- More shortwave UVR reaching the Earth
- Skin cancer
- Cataracts
- Eye damage leading to cloudy vision

Health impacts of global environmental change

Vector-borne diseases

Changes in temperature and rainfall regimes will alter the geographical distribution of optimal conditions for most vectors
Specifics: malaria, Zika, Lyme disease, West Nile virus

Agricultural productivity

1 Direct impacts
- Increased yields in mid and high latitudes
- Crops such as maize viable in lower latitudes
- Longer growing season in some specific areas, e.g. Russia
- Heat stress and water loss by evaporation in tropical areas
- Projections of increased rainfall in high latitudes and decrease in the tropics and sub-tropics

2 Indirect impacts
- Increased temperatures may make pests and disease more widespread
- Changes to water availability
- Sea level rise in conjunction with low-lying agricultural areas will decrease yields

Nutritional standards

1 Developed regions
Increased food prices may lower dietary quality
Climate change mitigation may increase consumption of foods whose production reduces greenhouse gas emissions

2 Developing regions
Some impacts where increased livestock production will increase methane, require more land for fodder production and more clearance of forests

3 Less developed regions
Rainfall unpredictability and drought will lead to lower crop yields, over-reliance on fewer crops and malnutrition

Figure 10.22 Health impacts of global environmental change

Prospects for the global population

Over the course of the twenty-first century world population is likely to rise by 50 per cent, in contrast to the 400 per cent of the last 100 years.

World population growth rate has fallen from 2 per cent to 1 per cent. The main drivers of world population growth are fertility rate and life expectancy:

+ **Fertility rate:** the UN forecasts a downward trend in global fertility rates from 2.5 to just over 2 children per woman during the twenty-first century.
+ **Life expectancy:** more people are living into old age as a result of medical advancements and improved standards of living.

Predictions of population growth are difficult to make and several scenarios have been put forward. Fertility rates in Sub-Saharan Africa, the growth of population in China and the success of education programmes for girls are three important contributing factors. Common views include the following:

+ World population growth will continue to slow to some extent.
+ Africa will show the greatest increases.
+ There will be continued population decline in parts of Europe and Japan.
+ Life expectancy will continue to rise.
+ Average age of the population will continue to rise.
+ Ageing populations will present new challenges.
+ Global fertility rates will continue to fall.
+ India may well overtake China as having the largest population size as it reaches its peak.

Now test yourself

TESTED

37 What are the main drivers of present-day world population growth?

38 Outline the challenges presented by an ageing population.

Answers on p. 271

Projected distributions

There is a degree of uncertainty, but a 2019 UN World Population Prospects report suggested that:

+ Africa is likely to account for more than half of global population growth from 2020 to 2050
+ Australia, New Zealand, central and southern Asia, Latin America and the Caribbean, eastern and Southeast Asia and Europe and North America will experience low rates of population growth
+ in 47 of the **least developed countries**, of which 33 are in Africa, population growth will remain high
+ nine countries, including India, Nigeria, Pakistan and the USA, will contribute more than half the world's population growth
+ most of Sub-Saharan Africa, parts of Asia and Latin America will be able to take advantage of the demographic dividend due to an increase in working-age population.

Exam tip

Remember that possible futures is an important concept in this unit. Prepare your ideas and take note of the difficulty in making projections about population change.

Critical appraisal of future population–environment relationships

There is a crucial debate as to whether **population** or **consumption** levels are the more significant threat to the environmental limits of the Earth.

The main challenge is to continue to supply the resources (food, water, energy) needed for survival in a sustainable manner.

It is feared that global population growth and the resulting increase in resource consumption will pose a major threat to the environment. Figure 10.23 summarises some of the threats and opportunities presented by population and resource pressures.

Check your understanding and progress at **www.hoddereducation.co.uk/myrevisionnotesdownloads**

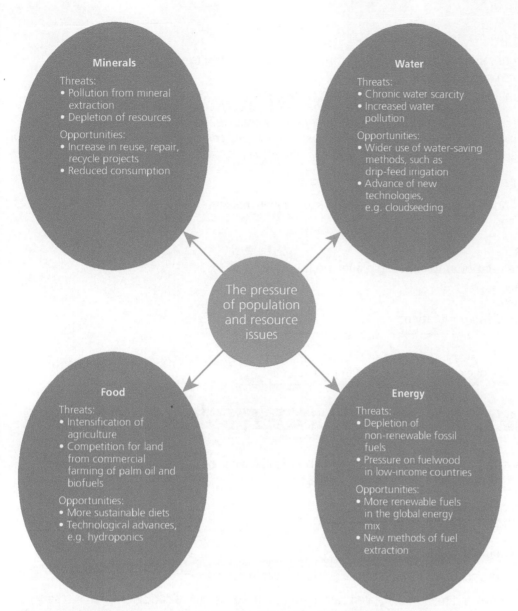

Figure 10.23 Threats and opportunities presented by population and resource pressures

Case studies

A country/society experiencing specific patterns of population change, for example Japan.

This case study should illustrate and analyse:

+ the character, scale, and patterns of change
+ relevant environmental and socio-economic factors
+ implications for the country/society.

What do I need to know?	Content and suggested revision methods
An overview of: + the character of population change + scale of population change + patterns of population change	The **character** of the population changes can be summarised in bulleted notes, for example: + Japan has a distinctive culture centred on hard work and family loyalty; this means that there are high levels of employment and large numbers of economically active people. + Japan is the third largest economy in the world, so healthcare provision is excellent. The **scale** and **pattern** of population change can be summarised in a series of detailed annotations to a copy or sketch of Japan's population pyramid. An example of an annotation would be: *Japan has a natural decrease and population is steadily declining. Birth rate: 7/1000; death rate: 11/100. In terms of scale, estimates predict a decline of 20% by 2050.* Highlight key facts to learn.

Relevant environmental and socio-economic factors	Summarise factors in a table. **Remember:** factors cannot just be descriptive; they need to be linked to population characteristics.

Environmental factors	Socio-economic factors
Intensive farming on alluvial plains and a strong fishing industry provides food security which raises life expectancy.	Women receive good education, have focused careers and fewer children, lowering the birth rate.

Implications for the country/ society	Summarise implications in bulleted points. Add a brief summary of some of the government responses.

A specified local area

To illustrate and analyse the relationship between place and health related to:

+ the physical environment
+ socio-economic character
+ experience and attitudes of the population.

> **Making links**
>
> This case study could be based on one of your place studies from Chapter 8.

What do I need to know?	Content and suggested revision methods
The place context How health and well-being is impacted by: + physical environment + socio-economic factors + the experience and attitudes of the population.	This case study can be summarised in the form of a report. 1 The 'place' context and the physical environment can be presented as an annotated sketch map. Include details regarding: a) geographical location and features b) key transport links c) facts about the local economy d) significant developments, e.g. regeneration schemes. 2 Two or three well-annotated data presentation graphs, charts or diagrams can be used to summarise the **socio-economic** factors/character. 　+ **Economic characteristics:** data on the local economy would include job opportunities, local industries, levels of employment. 　+ **Social characteristics:** include facts about health, education, population structure, income levels, cultural characteristics. Remember, factors need to be related to health. For example, low wages and high levels of unemployment lead to deprivation, health services lack funding as local tax yields are low. In addition, poverty is related to poor diets and higher rates of infant mortality. 3 **Experience and attitudes:** Construct a table that lists some key health indicators on one side and a comment on how these are affected by experience and attitudes on the other.

Health indicators	Experience and attitude
High proportion of obese children	Poor diet due to low income
	Lack of recreational space
	Lack of affordable recreation opportunities

It is useful to write a **brief conclusion** to the report which summarises the key points and puts forward some solutions. This may include local strategies already in place.

Exam practice

1 Explain the link between climate and human activities in one climatic type you have studied. [6]
2 Analyse the data shown in Table 10.12. [6]

Table 10.12 Average ecological footprint per person compared with productive biocapacity per person for six countries

Country	Biocapacity available per person (gha)	Ecological footprint per person, 2000 (gha)	Ecological footprint per person, 2016 (gha)	Biocapacity reserve or deficit per person, 2016 (gha)
Bangladesh	0.4	0.6	0.8	−0.4
Brazil	8.7	3.0	2.8	**+5.9**
Canada	15.1	9.0	7.7	**+7.4**
China	1.0	1.9	3.6	−2.6
India	0.4	0.9	1.2	−0.8
UK	1.1	5.7	4.4	−3.3

3 To what extent are Malthusian perspectives on population change relevant today? [9]
4 Assess the factors that might account for the patterns shown in Figure 10.1. [9]
5 To what extent have socio-economic factors impacted health in a local area you have studied? [9]

Answers and quick quizzes online

Exam skills

Opportunities to practise geographical skills within this topic include:
+ use of statistical data, for example census data and a range of statistics on the vital measures of population change
+ techniques such as frequency distributions, measurements of correlation and testing for the differences between data sets (Student's t-test and Mann-Whitney U test) can be used to interpret and manipulate data
+ map analysis – to determine physical and human characteristics affecting the pattern of population change and the reasons for it, for example choropleth maps
+ use of GIS data to compare population distribution with climatic zones, soil types and food supply, for example
+ fieldwork skills: data collection, recording, presentation and analysis
+ report writing and the use of core ICT skills based on the collection, analysis and presentation of secondary (and primary where appropriate) data.

Summary

+ This chapter centres on the impact of population on the environment and the ability of the environment to support growing populations.
+ You need a clear overview of the current trends in population distribution and the factors affecting it.
+ Climatic factors and soil types are key environmental determinants of food production. This relationship is illustrated through two examples of climate types and two zonal soils.
+ A range of strategies exists to ensure a country's food security. You should have an understanding of some of the different strategies involved.
+ Environmental, social and economic factors affect the incidence of disease. Environmental factors can be further divided into climate, topography, air quality and water quality. This is illustrated through the study of one biologically transmitted disease and one specified non-communicable disease.
+ Vital rates of population change must be learned and understood in terms of their impact on total population.
+ The DTM should be understood in a variety of contexts, illustrated by the examples of two contrasting countries. However, its relevance to demographic and economic trends in the developing world today should be questioned. Further concepts of dependency and age–sex population structure should be studied alongside the model.
+ Youthful and ageing populations present a range of challenges and benefits to a country.
+ The migration process has a range of social, economic and environmental causes and impacts for both host country and the country of origin.
+ The principles of population ecology should be understood in terms of the concepts of carrying capacity and ecological footprint. These concepts can be applied to a variety of scales.
+ The concept of 'optimum population' is relative and must be viewed critically.
+ Different models and perspectives relating to the relationship between the environment and population growth have been put forward. Develop a critical evaluation of the PRP model and the views of Malthus and those presenting alternative ideas.
+ Future projections of global population change are difficult to make and uncertain. The key prospects and projections put forward should be understood, together with a critical appraisal of their validity.

Resource development

Concept of a resource

A **resource** is any aspect of the natural environment that can be used to meet human needs. Examples include fossil fuels, water, wood and minerals. Resources:

+ have **economic value** and can be used to improve a country's wealth and further development
+ are **unevenly spread** across the world – some countries are resource rich, others resource poor. Some countries have an abundance of one resource, such as minerals, but a lack of another, for example water
+ can be transported and traded.

An important concept is **resource security** – the ability of a country (or indeed the whole world) to ensure a safe, reliable and sustainable flow of resources to maintain existing levels of development and allow future generations to advance.

Resource classifications

There is a wide range of resources which can be categorised to aid understanding. Although the term 'resource' often relates to **physical resources**, geographers also acknowledge **human resources** such as population and capital.

A distinction is also made between **stock resources**, which are non-renewable, and **flow resources**, which are renewable and can be replaced. Figure 11.1 shows how resources are classified.

> **Exam tip**
>
> Be clear on subject-specific terminology when reading exam questions. There are important differences between terms such as stock and flow resources, critical and continuous flow, reserves and resources, exploitation and evaluation, and consumption and supply.

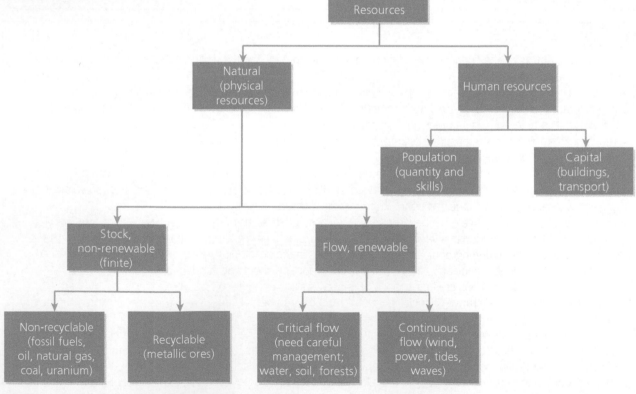

Figure 11.1 A classification of resources

Stock resource evaluation

Reserves are the part of a resource that it is economically, legally and technically viable to extract. **Resources** can be converted into reserves if there is technological advancement.

Mineral resources and reserves are further sub-divided into the following categories:

✚ **Measured reserves:** these can be estimated with confidence as quantity, grade and quality are well established. This results in a '**proven reserve**', which is economically viable to extract and has undergone a preliminary feasibility study.

✚ **Indicated reserves:** quantity, grade and quality can be estimated with less confidence than measured reserves and require further evaluation of the economic viability. The degree of confidence is less than with measured reserves but enough to allow a reliable estimate, which leads to reference to a '**probable reserve**'.

✚ **Inferred resources:** quantity, grade and quality can be estimated on the basis of only limited sampling. They remain '**possible reserves**' as there is insufficient information on tonnage or grade.

✚ **Possible resources:** there is knowledge of these resources based on the existence of other, mostly undiscovered deposits. They may become economically viable in the long term but there is less confidence about this than with inferred resources. This category consists of hypothetical and speculative resources.

Exam tip

In questions on the exhaustion of resources be clear on the categories of reserves and keep in mind that supply of resources fluctuates according to price and demand.

Natural resource development over time

Supply of resources depends on:

✚ physical risks – the quantity and quality of the resource, its location and accessibility
✚ geopolitical risks – trading confidence, possibility of conflict and the concentration of production in a small number of countries.

Only if the economic benefits outweigh the physical and geopolitical risks will production of the resource go ahead.

Figure 11.2 shows the sequence in which resource development takes place.

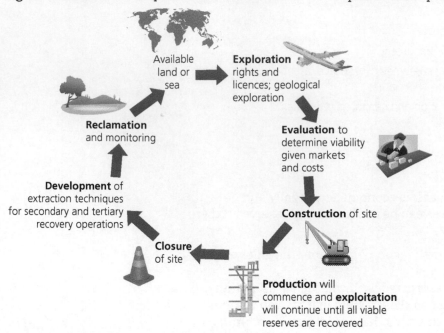

Figure 11.2 Resource development over time

Revision activity

It is important to understand the factors that affect resource development. Produce a revision table as outlined below to identify and explain those factors. Four factors are listed but you can add to these.

Factor	How it affects resource development
Demand	
Supply	
Technology	
Capital	

The concepts of a resource frontier and resource peak

Resource frontier

A resource frontier is an area where resources are brought into production for the first time. They exist at a variety of scales – local, between nations or cross continents.

In the core–periphery model (Friedman, 1963), resource frontiers exist within the periphery where there is a resource discovery prompting investment.

Resource peak

Resource peak is the time of maximum production of a reserve or of a resource as a whole. The theory of resource peak is represented in a 'bell-shaped curve' on a graph of production. Fossil fuel peaks vary as they depend on many factors.

> **Resource frontier** A newly colonised region where resources have been discovered and are brought into production for the first time.
>
> **Resource peak** The time of maximum production of a reserve or of a resource as a whole.

Sustainable resource development

The sustainability of resource development is a concern for several reasons:
+ As low-income countries develop, their resource use will increase.
+ The environmental impact of current levels of resource use is an issue.
+ There needs to be careful management of current resource use so that future generations can have access to the resources they require.

Check your understanding and progress at **www.hoddereducation.co.uk/myrevisionnotesdownloads**

Resource depletion is the use of resources faster than they can be replenished and it is a concept that mostly relates to non-renewable fossil fuels.

Table 11.1 summarises the two approaches to sustainable resource development: supply-side and demand-side management.

Table 11.1 Supply and demand strategies for sustainable resource development

Supply-side management	Demand-side management

Involves seeking methods of increasing the supply of resources:
+ increasing exploration efforts for existing non-renewable resources
+ increasing research efforts to develop:
  + more sustainable alternative or substitute resources to replace unsustainable ones
  + new technologies that are more sustainable and cause less environmental impact.
	Involves reducing consumption of resources, individually and at all other geographical scales:
	+ changing individual behaviour and lifestyle to discourage wasteful and/or extravagant use of resources
	+ developing technology to enable more efficient use of resources
	+ recycling after use
	+ reducing population growth with population control methods so there is less pressure on resources
	+ regulatory controls and frameworks as part of global governance, for example:
	  + Agenda 21
	  + Kyoto Protocol.

Additional approaches to sustainable resource development include:
+ attempts to minimise environmental impacts through advancing technology such as carbon capture and storage (CCS) technology
+ seeking alternative supplies of resources, particularly for energy.

Making links

See Chapter 7 page 152 for more on trading resources and page 149 on geopolitical issues.

Now test yourself TESTED ◯

8 Why is there concern over the sustainability of future resource development?

9 Explain the concept of a resource frontier.

Answers on p. 271

Environmental impact assessments

Environmental impact assessments (EIAs) can be used to evaluate the costs and benefits of resource development projects. They inform decisions by balancing economic gain with potential environmental impact and offer alternative approaches.

The EIA follows several stages:
+ An outline of the proposed development.
+ Description of the existing environment.
+ Assessment of the likely impact.
+ Outlining of mitigation.
+ The official publication of an environmental statement.
+ Decision for or against the proposal – there is a right to appeal from both sides.

Exam tip

If you are asked about the impact of resource development on the environment, keep in mind the classification of a) the physical environment and b) the human environment.

Exam tip

Keep in mind the difference between short-term present-day levels of demand and future possible long-term demands for resources.

Now test yourself

TESTED

10 Briefly outline the importance of sustainable resource development.

11 Why are EIAs important in relation to resource extraction?

Answers on p. 271

Natural resource issues

REVISED

Global patterns of production, consumption and trade of energy

+ Energy drives economic and social development.
+ There remains a marked energy gap between rich and poor nations.
+ A large proportion of the population in many low-income countries still have no electricity.
+ Developed countries consume nearly 70 per cent of the total supply of coal, oil and gas.

Global patterns of energy production

Key patterns of energy production:
+ The depletion of fossil fuel reserves in many traditional areas of production.
+ The development of fossil fuel reserves in Asia, Latin America and Africa.
+ New development of shale gas and oil reserves.
+ North America, the Middle East and Asia are the areas of highest energy production.
+ Energy-rich countries are led by China and the USA.
+ Exploration and production of gas and oil reserves in low-income countries have increased.

Figures 11.3–11.5 show the global patterns of coal, oil and natural gas production in 2017.

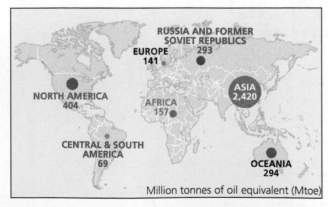

Figure 11.3 Global pattern of coal production, 2017

Exam tips

It is important to be able to account for patterns of energy production and consumption, for example link production to physical geography and cost of extraction and link consumption patterns to lifestyle and economic development.

Data on the patterns of energy production, consumption and trade needs to be up to date and show current trends.

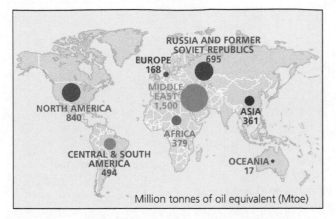

Figure 11.4 Global pattern of oil production, 2017

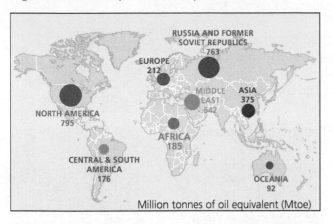

Figure 11.5 Global pattern of natural gas production, 2017

Global patterns of energy consumption

+ Globally, energy use is increasing by 2.5 per cent per year.
+ Energy usage per person has increased globally but varies between countries.
+ Energy consumption remains very high in China due to industrial demand and fuel consumption by private transport.
+ Energy consumption in the United States reached a record high of 2.3 Gt in 2018.
+ Energy consumption has decreased slightly (–1 per cent) in the EU due to energy-efficiency improvements.
+ Projections estimate a 50 per cent increase in global energy consumption between 2018 and 2050, mostly driven by Asia.

Consumption of coal

+ Coal consumption of oil is dominated by China, India, the USA and the EU.
+ These four account for 79 per cent of consumption.

Consumption of oil

+ The main consumer of oil is the USA, followed by the EU, China and the Middle East.
+ Oil accounts for 90 per cent of fuel energy but has been overtaken by gas and nuclear for electricity production.

Consumption of natural gas

+ The largest consumers of natural gas are the USA, the Middle East, the EU and Russia.
+ Gas now accounts globally for one fifth of energy consumption due to new discoveries, development of pipelines to transport gas and the fact that it is less polluting than coal.

Revision activity

Make revision notes to summarise the key features of the three maps in Figures 11.3, 11.4 and 11.5. A useful way of doing this is to annotate a world map as you will then have a visual impression of the patterns.

From your notes or further research, add a map for renewable and nuclear energy (these could be combined on one world map).

In your annotations:
a) describe the pattern of production
b) account for the pattern of production.

Exam tip

Oil is crucial to the global economy. The geopolitics of producers and consumers is extremely important. The dependence of growing economies on oil is illustrated by the position of the USA, Japan and Germany and by China's exploitation of energy supplies in Africa.

Now test yourself　　　　　　　　　　　TESTED ○

12 Why has energy use declined in the EU?

13 Outline one reason why Asia will drive future energy consumption.

Answers on p. 271

Global trade in energy

Energy security is a major goal for most countries. However, many, such as Japan, consume more than they can produce. Some countries that are resource-rich in energy have a surplus, while for some low-income countries this is a much-needed source of revenue.

+ **Oil:** this is the most traded energy resource because of demand for fuel. The largest net exporters of oil are Saudi Arabia, Russia and Kuwait. The largest net importers are the USA, China and Japan.
+ **Natural gas:** Russia is the largest exporter of natural gas, but export is more difficult because of the need for expensive pipelines. The main importers of natural gas are Japan, Germany and Italy.
+ **Coal:** trade in coal is less as it is low value and bulky. The UK and Germany import coal for electricity generation.

> **Energy security** The uninterrupted availability of energy sources at an affordable price.

> **Making links**
>
> See Chapter 7 page 148 for more on trade agreements and access to markets.

Figure 11.6 gives a summary of the changing patterns of energy consumption, production and trade.

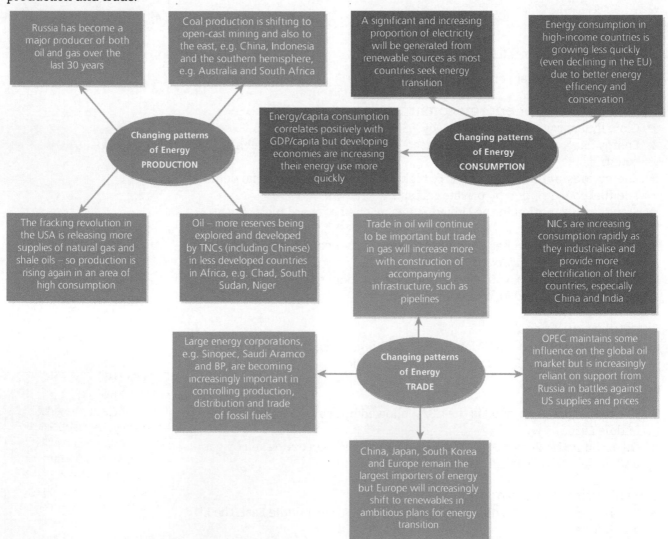

Figure 11.6 Changing patterns of energy production, consumption and trade

Now test yourself
TESTED ◯

14 State three factors that influence the growth in energy consumption.

15 Outline how energy security can vary across nations.

Answers on pp. 271–72

Global patterns of production, consumption and trade of mineral ores

Mineral ore production

Most developed countries have depleted their reserves of mineral ores. The shift in production has been to developing economies.

Table 11.2 lists the top producing countries of metal mineral ores in 2018.

Table 11.2 The top producing countries of metal mineral ores, 2018

Mineral ore	Uses	World total production (thousand metric tonnes)	Top four producing countries
Bauxite	Ore of aluminium	326,000	Australia, China, Guinea, Brazil
Iron ore	Steel	2,923,000	Australia, Brazil, China, India
Chromium	Alloys, stainless steel, electroplating	40,800	South Africa, India, Kazakhstan, Turkey
Copper	Alloys, pipes, wire	20,600	Chile, Peru, China, USA
Lead	Alloys, batteries	4,800	China, Australia, Peru, USA
Nickel	Alloys, stainless steel	2,233	Indonesia, Philippines, Russia, Norway, New Caledonia
Tin	Alloys, containers	352	China, Indonesia, Burma, Bolivia
Zinc	Steel alloys, galvanised metals	12,400	China, Peru, Australia, USA

There is increased production of metal ores in emerging economies:
+ South America, such as Brazil and Chile
+ South and Southeast Asia, such as India and Indonesia
+ Africa, such as Guinea and South Africa.

Much of this production is achieved through foreign investment and this can lead to a 'resource curse', which does not form a stable, long-term economic base.

Mineral ore consumption

Most increases in consumption are in aluminium (threefold expansion), copper and zinc (consumption has doubled) and ferrous metals used to produce steel.

The USA, Europe and Japan have now been overtaken as consumers by China, South Korea and India. The change represents deindustrialisation in some regions and rapid growth and industrialisation in China and India in particular.

Mineral ore trade

The pattern of trade in mineral ores reflects the changing nature of consumers and producers. The economic growth, resource wealth and population growth in China mean that it dominates trading patterns in mineral ores.

This can have global repercussions for world trade if economic growth slows in China.

Trade in mineral ores is affected by:
+ inelasticity of demand for specific minerals (demand does not change as much as price)
+ global recessions, which lead to falling demand and prices

235

+ technological change, which can increase supply and reduce price – unless demand also increases
+ environmental concerns, which encourage recycling and discourage new extraction.

Revision activity

Both China and TNCs have a significant impact on the trade of mineral ores. Make revision notes to summarise the following:
+ the impact of China on the production and consumption of mineral ores
+ the role of China in mineral ore trade
+ Chinese investment in the exploration and extraction of mineral ores
+ the role of TNCs in mineral ore distribution.

Global patterns of water availability and demand

There is enough water on the planet for 7 billion people; however, the supply is distributed unevenly. Poor management, pollution and waste also prevent equitable and plentiful supplies.

There is a correlation between areas of water shortages and areas of high population growth, such as Sub-Saharan Africa. Figure 11.7 summarises the imbalance of water availability.

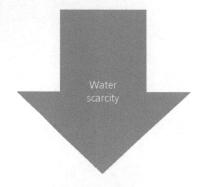

Physical water scarcity affects one fifth of the world's population.
Areas with <500 mm rainfall annually are known as having a water deficit, e.g. parts of Austraila, parts of North and Central America, central Asia and northern China.
Absolute water scarcity of <500 m³ per person exists in northern Africa (e.g. Algeria, Libya and Egypt) and the Middle East (e.g. Saudi Arabia).
Australia has physical water scarcity but can provide for the population due to effective management.
Economic water scarcity exists in Sub-Saharan Africa, parts of South America and southern Asia, where for economic reasons they cannot utilise water resources.
Water stress is where demand for water exceeds supply (<1,700 m³ per person) over a period of time, causing water shortages (e.g. India and parts of China).

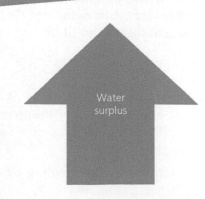

Temperate or tropical areas have plentiful rainfall, good runoff, lakes and aquifers (e.g. South America, North America, northern Europe and Southeast Asia) and therefore a **water surplus**.
Some countries have low rainfall but **efficient management** and therefore a surplus (e.g. the USA and Russia).
In water surplus areas there is usually efficient management of water quantity and quality
There is also **efficient usage** of water given the population.

Figure 11.7 Patterns of water availability

The pattern of demand for water

Figure 11.8 shows average renewable freshwater resources per person per capita water consumption by region.

Areas of the world most likely to suffer water shortages are:
+ those with rapid population growth (for example Sub-Saharan Africa)
+ those with a physical shortage of water due to climate conditions (for example North Africa).

Shortages will result in:
+ drought
+ more diseases resulting from contaminated water, for example cholera
+ over-exploitation of water
+ restrictions on economic development
+ food security issues.

Areas of the world most likely to have a water surplus include North and South America, northern Europe and Southeast Asia. This is due to factors such as:
+ plentiful rainfall
+ effective water management
+ low water usage (due to low population figures or technological advance).

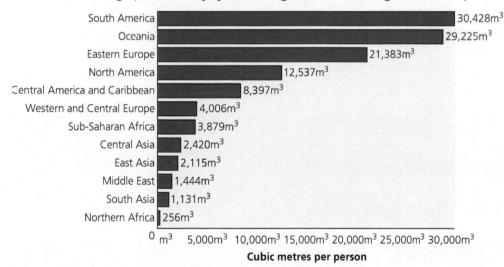

Figure 11.8 Average renewable freshwater resources per person, 2015

The geopolitics of energy, mineral ore and water resource distribution, trade and management

Table 11.3 outlines the geopolitics involved in energy, mineral ore and water.

Geopolitics The way in which geographical factors shape international politics.

Table 11.3 The geopolitics of energy, mineral ore and water resources

Energy	Mineral ore	Water

+ **OPEC**, an alliance of countries (e.g. Saudi Arabia and Venezuela) with oil surpluses.
+ Saudi Arabia has strong links with the USA and the West.
+ Arab Spring uprising since 2010 has led to conflict in many oil-producing countries, e.g. Libya, and oil supplies controlled by extremist militant groups.
+ The end of economic sanctions against Iran has improved its relationship with the West.
+ Russia holds vast energy reserves in gas in particular, on which many countries depend. Russia's involvement in conflicts with the Ukraine and Syria has put political relations under renewed strain.
	+ Dependence of Europe, China and Southeast Asia on mineral supplies from countries such as those in South America and Africa.
	+ There is a largely one-direction flow of trade in mineral ores from developing to developed countries.
	+ Some countries have limited the supply of ores, e.g. bauxite from Indonesia, in an attempt to raise prices.
	+ China has a leading role in the geopolitics of iron, steel and copper industries – it has invested heavily in mining operations (e.g. copper) in Africa. It produces large amounts of cheap steel which have flooded markets in countries such as the USA and the UK, leading to closure of plants and unemployment, e.g. Tata Steel sold its UK operations.
	+ A number of TNCs control mining operations, leading to social, economic and environmental responsibility (Rio Tinto, BHP Billiton).
		+ Water resources are shared by different countries, e.g. 276 transboundary river basins, 200 transboundary aquifers.
		+ Water conflict hotspots include:
		  + Nile – Egypt, Ethiopia and Sudan
		  + Tigris-Euphrates – Turkey, Iraq, Syria
		  + Aral Sea – Uzbekistan, Turkmenistan and Kazakhstan
		+ Potential areas of hydro-conflict include the Mekong, Ganges, Zambezi, Colorado and La Plata basins.
		+ Attempts to manage conflict hot spots include the Indus River Commission and the Berlin Rules on Water resources.

Water security

REVISED

Components of demand

Figure 11.9 shows end use of water in different parts of the world based on the three components of water demand:

+ **agricultural** (crop irrigation and livestock care)
+ **industrial** (water used as a coolant, for heating steam turbines or various processing operations such as textiles and food processing)
+ **domestic** (household and public/municipal).

OPEC Organization of the Petroleum Exporting Countries.

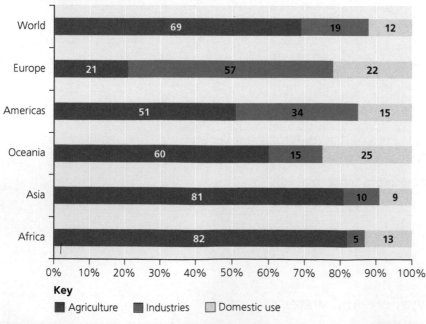

Figure 11.9 Variations in end use of water in different parts of the world

> **Making links**
>
> See Chapter 7 for more on geopolitics in global governance and global systems.

Check your understanding and progress at **www.hoddereducation.co.uk/myrevisionnotesdownloads**

Water stress

This is an imbalance of water use and available supply – the measure is availability of <1700 m³ per person per year.

Water stress affects the quantity and quality of water supplies.

Relationships of water supply to key aspects of physical geography

Exam tip

Remember that there are also relationships of water supply to human activity – importantly, the efficient use and effective management of water resources.

Climate

The two main components of climate that affect water supply are precipitation and temperature; seasonal changes will also have an impact.

Precipitation

Global **precipitation** is estimated to be on average 860 mm per year – 77 per cent falls over oceans and 23 per cent on land. For land-based fresh water supplies to be maintained there needs to be reliable and adequate annual rainfall; this may be seasonal but storage can level out these differences.

Temperature

+ High **temperatures** lead to high rates of evaporation, which can mean that inputs of precipitation are lost to the atmosphere before they impact on water supply.
+ Where the climate is cooler and less windy there will be lower rates of evaporation.
+ Where climate is characterised by very low (sub-zero) temperatures there will be freezing for long periods of time and fresh water supplies will be inaccessible until a seasonal period of thawing.

Seasonal changes

Seasonal changes in the balance between precipitation and **evapotranspiration** will also affect underground water supplies.

In winter, when there is more precipitation, this will lead to a surplus and rising levels of underground water sources; the reverse will be true in summer.

Geology

Geology affects the location of reservoirs. Impermeable rock will ensure that water is not lost due to seepage and rocks must be stable, as faulting and movements will damage or destroy dams.

Drainage

+ The **drainage basin** is the area drained by a river and its tributaries.
+ It receives inputs of precipitation and the outputs include channel runoff, evapotranspiration and groundwater flow.
+ The operation of the drainage basin system is determined by local physical factors such as climate and geology.
+ The relationship between the inputs and outputs will determine the water supply.

Making links

See Chapter 1 page 13 for more on the drainage basin system.

Strategies to increase water supply

Catchment Abstraction Management Strategies (CAMS)

CAMS are used in the UK to ensure water resource sustainability. Water abstractions over 20 m³ per day require an abstraction licence; whether a licence is granted or not depends on 'the amount of water available after the needs of the environment and existing abstractors are met and whether the justification for the abstraction is reasonable' (Environment Agency, 2013).

River diversion

River diversion transfers water from a catchment with a water surplus to one of water shortage. Transfer can take place by aqueducts, river diversion, canal or pipeline. For example:
+ transfer from sparsely populated mid Wales to the Midlands conurbation
+ in the Sindh province of Pakistan where irrigation channels transfer water from the Indus basin.

Reservoirs

Reservoirs are a method of **water storage** for winter surpluses of water and they also regulate river flow. The water is released via pipeline to the public supply. Reservoirs may be multi-purpose, such as the Three Gorges Dam in China, which is used to generate electricity and promote tourism.

Desalination

Desalination is the removal of salt from seawater; however, it is expensive and a significant source of greenhouse gas emissions. The two main methods are:
+ reverse osmosis (filtering of seawater at high pressure)
+ distillation (water is boiled, the steam condensed and collected, salt is left behind).

Check your understanding and progress at **www.hoddereducation.co.uk/myrevisionnotesdownloads**

Because of the expense and technology required it is confined to more wealthy countries with water security issues, such as Saudi Arabia, the USA and Australia.

> **Exam tip**
>
> Although positive environmental impacts are less common, for balance attempt to identify any that are relevant, for example the Three Gorges Dam: reforestation on surrounding steep slopes; increasing the water supply may relieve damaging environmental pressures on water extraction elsewhere.

Now test yourself   TESTED ◯

22 State two advantages and two disadvantages of reservoirs as a means of water storage.

23 Explain the process of desalinisation.

Answers on p. 272

Environmental impacts of a major water supply scheme

See the revision activity.

Revision activity

In class you will have studied an example of a major water supply scheme. Make summary revision notes using the outline in Figure 11.10 which is based on the Aswan High Dam. Colour-code positive and negative environmental impacts on your revision summary.

Remember the need for some place-specific facts and that the focus of the specification is on environmental impacts of a major dam and/or barrage and its associated networks. Common examples include the Aswan High Dam, the Three Gorges Dam, the Thames or the Tees barrage.

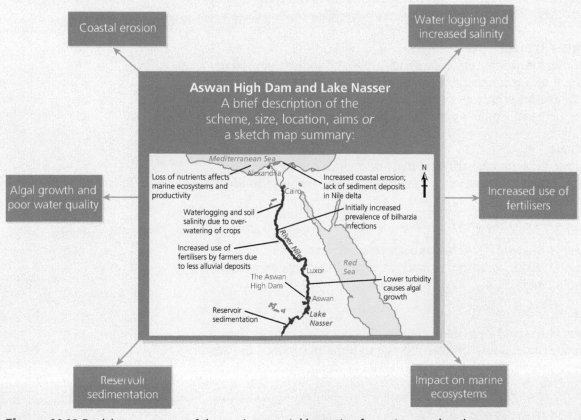

Figure 11.10 Revision summary of the environmental impacts of a water supply scheme

My Revision Notes: AQA A-level Geography Second Edition

Strategies to manage water consumption

Strategies to reduce domestic wastage include:

+ installation of water meters
+ low-flush and dual-flush toilets
+ encouraging the use of showers and not baths
+ new technology to save water in domestic appliances – short wash cycles for dishwashers and washing machines
+ collecting rainwater to use in gardens
+ push-button taps and tap restrictors.

Strategies to **reduce agricultural use** include:

+ drip-feed irrigation, which delivers water to the plant base
+ rain sensors, which shut down irrigation systems when it rains
+ digging boreholes to water-bearing rock
+ contour ploughing to reduce runoff
+ increasing the percentage of organic content in soils as this retains water
+ collection and recycling of water.

Strategies to **reduce industrial use** include:

+ industries' use of all the attempts outlined above to conserve domestic water supplies in addition to educating the workforce on the need to conserve water
+ looking at the many examples of industries that have implemented water conservation measures, such as Walkers crisps using water meters and recycling systems.

> **Exam tip**
>
> Remember that there will be different approaches to water conservation in developed and developing countries.

Sustainability issues associated with water management

Virtual water trade

The concept of virtual water relates to agricultural and industrial products. The requirement of water needed is estimated and the product has a virtual water value. Countries with water scarcity import foods that require irrigation and therefore there is a transfer of virtual water.

Conservation

Afforestation projects increase interception and reduce runoff, thereby retaining water in the drainage basin system and maintaining groundwater stores at a sustainable level.

Recycling

Recycled water is waste water that has been treated/purified so that it can be reused for non-drinking purposes, such as agricultural irrigation and watering of public parks.

'Grey water'

Grey water has been used for washing or cleaning but has not been in contact with faecal matter, so it does not need sewage treatment. It can be used for flushing toilets, gardening or irrigation. It can be used widely in offices, hotels and leisure centres.

Groundwater management

Aquifers can be recharged artificially by pumping water underground and diverting rivers and/or storm water to permeable surfaces.

> **Making links**
>
> See Chapter 1 page 11 for more on the water cycle.

Water conflicts at a variety of scales

Local

In northern Chile there is conflict over the use of scarce water resources. Since the discovery of copper reserves, mining companies that are more affluent have paid for water to be diverted away from vineyards and into copper mines.

National

Within countries there may be disagreement over the location and use of water resources. For example:

+ UK: the West Midlands conurbation receives much of its water from the Elan Valley reservoir scheme in North Wales. In the past this has created concern over the environmental impacts felt in Wales as a result of supply water installations for Birmingham. New investment in the water supply for the West Midlands is taking place in Worcestershire.
+ Southern Spain: there is disagreement between farmers and local councils over the use of large quantities of water on golf courses.

In poorer countries where tourism provides much-needed income there is also concern over the supply of water to large hotels, which could be directed to shanty towns and impoverished urban communities.

International

The Tigris and Euphrates rivers both have their source in Turkey. The rivers then flow through Syria and Iraq to the Persian Gulf. This has caused deep conflict in the area as all three countries require the water from these rivers but Turkey is in a position to control the flow.

Syria has the Tabaqah Dam, which forms Lake Assad, and Turkey has the South-eastern Anatolia Project, which involved building 21 dams on rivers in Turkey.

In Iraq, which is further downstream, 50 per cent of farmers in the south of the country have moved to occupations in urban areas because of the water shortages in agriculture.

> **Making links**
>
> See Chapter 7 for more on geopolitical issues.

> **Now test yourself** TESTED ◯
>
> 24 Define a) virtual water, b) grey water.
>
> 25 Why is there international conflict over domestic water resources?
>
> **Answers on p. 272**

Energy security REVISED ◯

Global energy supply faces the following challenges:

+ increasing transition from fossil fuels to cleaner renewable energy
+ increased risk of disruption to energy supply due to political tension
+ the persistence of energy poverty in developing countries.

Sources of energy

Primary energy sources are obtained in their natural form.

Secondary energy sources are sources that have been converted from primary sources into manufactured sources – mainly heat, electricity or fuel.

Fossil fuels

Coal

Coal supplies 26 per cent of global energy needs and 37 per cent of global electricity production. Most is consumed domestically in the country of production.

243

Costs of mining are influenced by depth, thickness and quality of coal. The main types of extraction are open cast at the surface or deep mining. There are concerns over greenhouse gas emissions.

Oil

Oil supplies 30 per cent of global energy as petrol, oil and diesel.

Many developed countries cannot satisfy their oil requirements and import oil, which can create geopolitical tensions.

Transport costs via oil tankers are relatively low and it is a flexible method of transport. However, there are environmental concerns over leakage.

Natural gas

Natural gas accounts for 22 per cent of the global supply of energy and is a growing source of fuel as it reduces dependency on oil and has fewer environmental concerns. However, extension of pipelines into more remote and politically unstable areas presents future problems.

There is an increase in proven reserves from shale gas and there have been high levels of investment in gas transport and distribution. Of the transported gas, 21 per cent is in the form of liquefied natural gas.

Nuclear energy

Nuclear power supplies more than 6 per cent of global energy. However, it requires high levels of investment, technology and careful handling of waste, and as a result most production is in the developed world.

Opinion over this fuel source is divided due to the potential for accidents (Three Mile Island, 1979; Chernobyl, 1986), and radioactive materials remain toxic for thousands of years. Nuclear power plants are a potential target for terrorism, and it is difficult to find suitable sites.

Advantages include the fact that nuclear power is efficient, holds large reserves and causes less pollution than fossil fuels.

Solar energy

Solar energy is produced via 'passive' solar architecture, such as south-facing windows, by boiling water to produce steam to drive turbines or by using photovoltaic (PV) cells. There are many localised sites but the main sources are in desert environments where there are clear skies for at least 300 days a year. There is huge potential in solar energy.

Wind energy

The potential of wind energy is determined by wind speed, and spacing and size of turbines. Wind farms connect to the electricity grid network. Sites with reliable prevailing winds and long coastlines are ideal locations.

Wind turbines can be visually intrusive and many of the best sites have environmental status, but offshore wind farms are a way round this.

Biomass energy

Biomass energy accounts for 10 per cent of global energy, mainly in developing countries. Wood is the most important biomass fuel and providing there is careful management and adequate replacement timescales, wood is a sustainable fuel source. Other biomass fuels include animal and vegetable waste.

Biomass can also be burned in thermal power stations. Carbon neutrality is questionable as a carbon sink is lost.

Hydroelectric power

Hydroelectric power provides 2 per cent of the global energy supply. Large-scale projects include the Three Gorges Dam in China and the Itaipu Dam in South America.

> **Making links**
>
> There is an important link between energy sources and physical geography. See more on page 246.

Check your understanding and progress at **www.hoddereducation.co.uk/myrevisionnotesdownloads**

This source requires powerful rivers with high annual discharge, steep gradients and natural storage. It is a popular alternative source of fuel as it is renewable, carbon free and can generate large amounts of electricity.

However, it needs huge capital outlay and large projects have a range of environmental, social and economic impacts.

Tidal power

Tidal power has huge potential but only a small amount is used at the moment. Ideal sites are in estuaries with large tidal ranges and where barrages can be built at relatively low cost.

Problems include large capital cost and environmental issues concerning the impacts on mudflats and salt marshes, and the disruption to natural tidal flows.

Wave power

Further development is required to make wave power commercially viable. Technology includes the Pelamis Wave Energy Converter and the LIMPET device.

There is a regular supply of wave power and it is pollution free, but transmission from offshore locations presents problems.

Geothermal energy

Geothermal energy makes use of the temperature rises within the Earth's crust. The most common way of capturing this energy source is to tap into hydrothermal convection systems. Cool water is pumped into the Earth's crust, it heats, rises, steam is captured and is used to drive electricity generators.

This power source is clean and sustainable but reliant on suitable sites.

Revision activity

On a piece of A3 paper, create a table to summarise the main features of different energy sources. Organise your table into three sections. Coal has been done for you.

Energy source	Main demand and supply issues	Advantages	Disadvantages
Coal	Supplies 26% of global needs.	Advancing technology could mitigate pollution concerns.	Concern over greenhouse gas emissions.

Now test yourself　　　　　　　　　　　TESTED ◯

26 What factors affect the cost of coal mining?

27 Why does oil supply create geopolitical tensions?

28 Why is gas seen as the most stable fossil fuel source for the future?

29 What are the advantages and disadvantages of nuclear power?

Answers on p. 272

Components of demand and energy mixes in contrasting settings

Components of demand refer to the way energy is used – industrial, commercial, domestic and transport are the most common categories.

Energy mix is the composition of different sources from which a named area obtains its energy. This may be a country or globally. Several factors will influence a country's energy mix:

My Revision Notes: AQA A-level Geography Second Edition

- availability of energy resources within the country
- inertia (retaining a certain mix due to the costs of changing)
- government policy
- geopolitics
- level of development
- physical/locational conditions
- the government's wish to diversify.

Contrasting settings refer to an overview of how countries in different contexts obtain their energy. The contrast may be in the type of energy source – fossil fuel or renewable – the level of economic development or the indigenous supplies available.

Table 11.4 contrasts the energy mix in Iceland and France. Other useful contrasting countries include Brazil (with a high dependency on sustainable energy supplies) and Mali (a low-income country dependent on biomass).

> **Exam tip**
>
> Learn exact percentage figures in relation to a country's energy mix and remember the four key factors in determining the mix: available, affordable, reliable and clean.

Table 11.4 Energy mix in contrasting settings: Iceland and France

Iceland	France
Energy mix: HEP 21%, geothermal 67%, fossil fuels 12%	Energy mix: nuclear 71.7%, renewable (geothermal/solar/wind) 21.3%, fossil fuels 7%
• Very low on fossil fuel use. • Abundant geothermal energy due to positioning on a constructive plate margin. • Huge potential for HEP from fast-flowing rivers with steep gradients. • As a small affluent country it can afford the technology needed for both geothermal and HEP.	• High usage of nuclear fuel – 59 power stations, supplying nearly 80% of electricity. • Many large rivers for cooling purposes in nuclear fuel production. • Government policy to improve energy security after the 1970 OPEC crises. • Has few fossil fuels.

Now test yourself TESTED ◯

30 Explain the term 'inertia' in the context of a country's energy mix.

31 Why do governments seek to diversify their energy mix?

Answers on p. 272

> **Exam tip**
>
> Be ready to evaluate the **relative importance** of different factors influencing a country's energy mix.

Relationship of energy supply to key aspects of physical geography

Table 11.5 shows the links between physical geography and energy supply.

Table 11.5 How physical geography plays a part in energy supply

Climate	Geology	Drainage
• High levels of sunlight can be harnessed for solar energy. • Best location for solar energy is in tropical and subtropical areas. Sun is higher in the sky so potential is maximised. • Solar energy has more potential in mountainous areas where the air is thinner and sunlight is scattered less. • Wind is a source of energy but can be intermittent. • Minimum wind speed required is 7–10 mph. In high winds of 50–80 mph most turbines will shut down. • Wind: high-density air provides more energy so lower altitudes and cooler air are most effective.	• Coal – formed from plant debris buried under layers of mud and sand; heat and pressure from subsequent layers gives rise to a process of coalification. Types of coal include anthracite, bituminous and lignite. • Oil and gas – these are hydrocarbons of organic origin. Settling of dead plants and animals at the bottom of the sea led to fossilisation in sedimentary source rocks. • Sedimentary rock formations hold oil and gas or both within their pores. • New technology has allowed the extraction of oil and gas from shale rock by a process of fracking, where water and chemicals are pumped in at pressure to fracture the rock and liberate the oil and gas from the pore spaces.	• Size and shape of drainage basins influence the potential for dam building and HEP. • Vital factors for HEP are the volume of water that can be captured – the flow. • The other vital factor for HEP is the height the water will fall – the head. • Dam building is expensive, so a long, narrow, steep-sided valley basin is most suitable.

Check your understanding and progress at **www.hoddereducation.co.uk/myrevisionnotesdownloads**

Energy supplies in a globalising world

Competing national interests

As more countries progress along the development continuum, there will inevitably be increased competition for declining supplies of non-renewable energy. International competition will centre on:

+ continued dependence of Europe and other western economies on Middle East oil
+ growing dependence of Asia on oil
+ China's continued investment in and exploration of energy supplies in Africa
+ natural gas continuing to be a fast-growing sector.

Role of TNCs

Powerful TNCs – Royal Dutch Shell, BP and Exxon-Mobil – dominate the oil trade. Due to their involvement in exploration, production and distribution, these TNCs will continue to have significant influence over governments and the spread of economic activities.

Environmental impacts of a major energy resource development

Revision activity

In class you will have studied an example of a major energy resource development. Make summary revision notes using the outline in Figure 11.11. Colour-code positive and negative environmental impacts on your revision summary. Your example may be based on local fieldwork.

Remember the need for some place-specific facts and that the focus of the specification is on environmental impacts of a major energy resource development such as an oil, coal or gas field and associated distribution networks.

Your example may be based on local fieldwork.

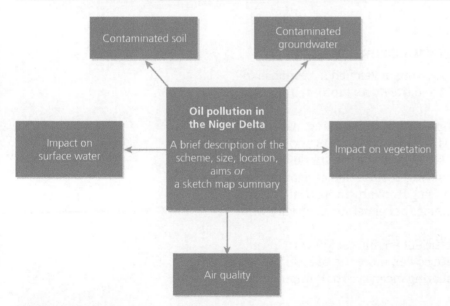

Figure 11.11 Revision summary of the environmental impacts of a major energy resource development

Making links

See Chapter 7 page 154 for more on the role of TNCs.

Exam tips

Although positive environmental impacts of energy production are less common, for balance attempt to identify any that are relevant.

A useful way of looking at the environmental impact of energy production is to consider the scale, permanence and location of the project.

Strategies to increase energy supply

Oil and gas exploration

Countries have been more willing to grant exploration rights to TNCs to find new reserves of oil and gas or investigate speculative reserves. Price rises in oil and gas also prompt TNCs to invest in technology to advance exploration.

An example of this new exploration is the award of licences to fracking companies such as Cuadrilla in northwest England and the East Midlands.

Nuclear power

Many people consider nuclear power to offer the most effective long-term solution to over-reliance on fossil fuels.

In 2007 in the UK, the Energy Policy Review announced that more nuclear power stations would be built, with some on existing sites at Hinkley Point (Somerset) and Wylfa (Anglesey).

Development of renewable sources

Financial support has been offered to renewable energy schemes, such as the Renewable Obligations and Feed-in Tariff schemes. These schemes have seen rapid growth in wind, solar and biomass energy sources in the UK.

Now test yourself   TESTED

32 How do TNCs influence oil supplies?

Answer on p. 272

Strategies to manage energy consumption

Strategies to manage energy consumption are developed in response to international plans, such as Agenda 21 and the Kyoto Protocol. Efforts to conserve energy include:

+ **household energy saving:** draught-proofing, new building materials that reduce heat loss, use of solar panels and biomass boilers and improved designs for passive solar heating – large, south-facing windows
+ **industrial and commercial energy saving:** installation of heat-recovery systems to collect and reuse heat arising from any industrial process, also combined heat and power systems that generate electricity while capturing usable heat
+ **energy saving in transport:** more efficient engine design with lower emissions, hybrid cars using electric power, more use of bioethanol fuel, road tax based on emissions, car-sharing incentives and improvements in public transport.

Exam tip

Remember that renewable power will not replace fossil fuels and nuclear power. Even by 2020 few countries will have as much as 25 per cent of their electricity generated by renewable resources.

Check your understanding and progress at **www.hoddereducation.co.uk/myrevisionnotesdownloads**

Sustainability issues associated with energy production

Acid rain

The issues associated with acid rain are outlined in Figure 11.12.

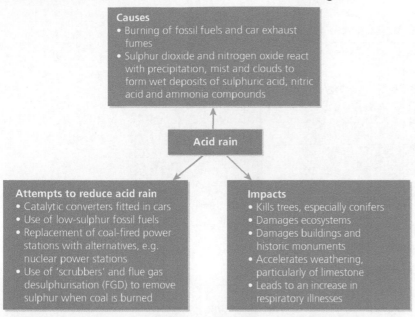

Causes
- Burning of fossil fuels and car exhaust fumes
- Sulphur dioxide and nitrogen oxide react with precipitation, mist and clouds to form wet deposits of sulphuric acid, nitric acid and ammonia compounds

Acid rain

Attempts to reduce acid rain
- Catalytic converters fitted in cars
- Use of low-sulphur fossil fuels
- Replacement of coal-fired power stations with alternatives, e.g. nuclear power stations
- Use of 'scrubbers' and flue gas desulphurisation (FGD) to remove sulphur when coal is burned

Impacts
- Kills trees, especially conifers
- Damages ecosystems
- Damages buildings and historic monuments
- Accelerates weathering, particularly of limestone
- Leads to an increase in respiratory illnesses

Figure 11.12 The causes, impacts and reduction of acid rain

The enhanced greenhouse effect

Due to the Earth's natural greenhouse effect, solar insulation is trapped in the lower atmosphere. The rise in the combustion of fossil fuels over the past 200 years has increased the amount of greenhouse gases, enhancing this effect.

There is growing agreement among scientists that the CO_2, methane and nitrous oxides released by burning fossil fuels are leading to climate change.

> **Exam tip**
>
> Do not confuse renewable and sustainable energy – even biomass fuels must be used in a sustainable manner.

Nuclear waste

Nuclear waste remains highly radioactive for thousands of years and consequently it has to be disposed of safely. Spent fuel rods and fission products are removed from nuclear reactors, vitrified into solid blocks and stored in lead-lined containers underground.

+ Burial sites with depths of 200–1000 m need to be found in areas of geological stability.
+ It is expensive to purchase land for the burial sites.
+ Transport to the burial sites needs to be safe and is also costly.
+ There are political concerns regarding the sites posing a target for terrorism as well as social concerns from local pressure groups.

Energy conservation

See the revision activity.

> **Revision activity**
>
> Following the example for acid rain (Figure 11.12), draw two more diagrams to summarise the causes, impacts and reduction of:
>
> a) the enhanced greenhouse effect
>
> b) nuclear waste management.

> **Revision activity**
>
> From your own notes or additional research, make revision notes on a range of attempts to conserve energy in:
>
> + the home
> + businesses
> + transport systems.
>
> You can use a simple table to organise your notes.

My Revision Notes: AQA A-level Geography Second Edition

Mineral security

The rapid growth in metal and mineral use has widespread implications for the security of future reserves and the environment.

Making links

See Chapter 7 for more on resources in the global commons – Antarctica.

Figure 11.13 details various facts about copper.

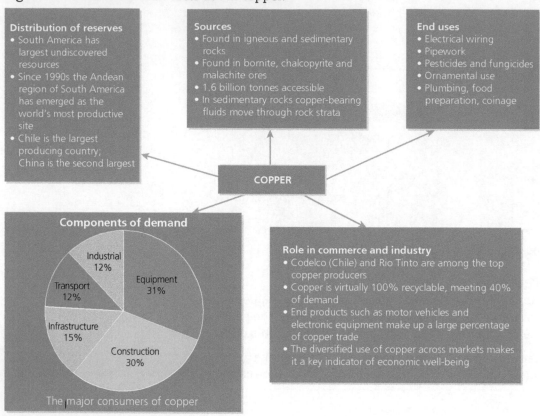

Distribution of reserves
- South America has largest undiscovered resources
- Since 1990s the Andean region of South America has emerged as the world's most productive site
- Chile is the largest producing country; China is the second largest

Sources
- Found in igneous and sedimentary rocks
- Found in bornite, chalcopyrite and malachite ores
- 1.6 billion tonnes accessible
- In sedimentary rocks copper-bearing fluids move through rock strata

End uses
- Electrical wiring
- Pipework
- Pesticides and fungicides
- Ornamental use
- Plumbing, food preparation, coinage

COPPER

Components of demand

Industrial 12%
Equipment 31%
Transport 12%
Infrastructure 15%
Construction 30%

The major consumers of copper

Role in commerce and industry
- Codelco (Chile) and Rio Tinto are among the top copper producers
- Copper is virtually 100% recyclable, meeting 40% of demand
- End products such as motor vehicles and electronic equipment make up a large percentage of copper trade
- The diversified use of copper across markets makes it a key indicator of economic well-being

Figure 11.13 Mineral security example: copper

Key aspects of physical geography

The key aspects of physical geography associated with the occurrence of copper are outlined in Table 11.6.

Table 11.6 Copper: geological conditions and location

Geological conditions	Location
Magmatic deposits: linked to magma; crystallisation produces ores containing nickel–copper deposits and platinum metals.	Advances in technology have allowed even the most difficult and remote sources to be explored (deserts, e.g. the Atacama Desert, and forests, e.g. the tropical rainforest of Brazil).
Hydrothermal deposits: hot solutions containing dissolved minerals flow into rock cracks and fissures and move towards the surface where they cool – copper, lead and tin form in this way.	The shape, size, quality and grade of the ore deposit will determine whether it is mined underground or open cast.
Metamorphogenic deposits: formed by intense heat and pressure over a long period of time – iron ore, gold and uranium are formed in this way.	
Sedimentary deposits: contain copper, lead and zinc.	

Check your understanding and progress at **www.hoddereducation.co.uk/myrevisionnotesdownloads**

Environmental impacts of a major mineral resource extraction scheme and associated distribution networks

See the revision activity.

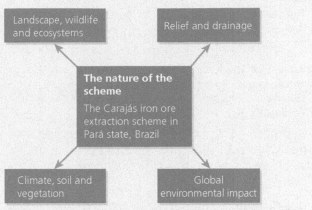

Figure 11.14 Revision summary of the environmental impacts of a major mineral resource extraction scheme

Sustainability issues associated with ore extraction, trade and processing

Ore extraction

Table 11.7 outlines the impacts of ore extraction and measures taken to improve sustainability.

Table 11.7 Impact of ore extraction and measures to improve sustainability

Impact	Measure to improve sustainability
Habitat loss	Restoration plans, e.g. Carajás project, Brazil
Noise and dust pollution	Baffle mounds and water sprays
Toxic leachates	Passing mine drainage water through a filter of limestone to immobilise toxic metals

Trade

In 2009, the World Economic Forum launched the Mining and Metals Scenarios to 2030 project. It examined the economic, trade and geopolitical future of the minerals and mining sectors. It presented three future scenarios:

+ Green Trade Alliance (GTA) – environmental standards used as protectionist measures in trade of ores and metals
+ rebased globalism – free-market principles upheld to enable poorer nations to avoid the 'resource curse'
+ resource security – based on national self-interest, protectionist measures which limit cross-border flows of resources.

Processing

Recycling of metals and minerals has been established in many countries as a result of Agenda 21. Recycling rates are good and well organised for some materials. Global figures for recycling include:

+ 70–90 per cent of iron and steel
+ 40–50 per cent of copper
+ 90 per cent of tin.

Media campaigns, legislation and financial incentives all help recycling. Mixed materials, transport and labour costs can hinder the process.

> **Exam tip**
>
> Although positive environmental impacts are less common, for balance attempt to identify any that are relevant, for example attempts to landscape and restore the site post extraction.

> **Now test yourself**
>
> 33 Why is copper used as an indicator of economic well-being?
>
> 34 Explain how ore extraction is damaging the environment.
>
> **Answers on p. 272**
>
> TESTED ⬤

251

Resource futures

Alternative energy, water and mineral ore futures and their relationship with a range of technological, economic, environmental and political developments are outlined in Table 11.8.

> **Exam tip**
>
> Possible futures is an important part of the specification, so make sure that you are able to put forward your ideas and evaluate the alternatives.

Table 11.8 The future of energy resources, water resources and mineral ores

The future of energy resources			
Technology	**Economic**	**Environmental**	**Political**
Hydrogen – a high-energy fuel which could replace fossil fuels. Large investment in this by TNCs. **Gasification of coal** – coal is converted to gas in deep, inaccessible resources. **Nuclear** – advances have been made in safer technology to make nuclear fuel a more viable alternative.	Energy flows towards Asian markets will continue to gather pace. By 2020s, non-OPEC oil supply from Canada, Brazil and the USA may decline and there will be more reliance on the Middle East. Natural gas production will increase outside of Europe.	**Carbon capture and sequestration** will eliminate carbon emissions from fossil fuels. On a national scale there is likely to be a move towards smaller power stations and energy from waste alternatives.	The influence of TNCs will continue to grow. Developing countries will rely increasingly on TNCs to develop their resources. As fossil fuels become depleted, there will be renewed pressure on remaining reserves, e.g. the Arctic, which is less well protected than Antarctica.

The future of water resources			
Technology	**Economic**	**Environmental**	**Political**
Osmotic distillation may be an alternative for ocean supply. Salt water greenhouse technology is well suited to arid parts of the world. Smaller-scale appropriate technology, e.g. water cones, can distil seawater in small quantities.	**Water shipping**, **water management** and **virtual water trade** will progress distribution of water resources.	**Integrated river basin management** will be needed in the future to mitigate the impacts of climate change on water resources.	Conflict hotspots will require close monitoring. The UN has a role to play in the future in: + reducing deaths caused by water-borne diseases + managing the release of chemicals and waste into water sources + conservation of water-related wildlife and ecosystems.

The future of mineral ores			
Technology	**Economic**	**Environmental**	**Political**
Remote sensing will enable large areas to be surveyed for new resources. Magnetometry will identify iron ore deposits. Seismic surveys are increasingly used in underwater surveys.	Reserves of exploitable minerals are limited. High extraction costs and limited technology may prevent possible reserves being exploited. Prices are therefore likely to rise.	Exploration and mining will continue to put pressure on the environment. The economic gains of mining mineral resources may outweigh environmental concerns in emerging economies that are resource-rich, e.g. Latin America (Brazil, Peru) and Africa (Zambia, DR Congo). Environmental concerns specifically relate to reserves in Antarctica, Alaska and the oceans.	Governments have a continued duty to control mining operations and protect the environment. Compulsory EIAs will be a necessary step to ensure environmental protection.

Check your understanding and progress at **www.hoddereducation.co.uk/myrevisionnotesdownloads**

Case studies

A case study of <u>either</u> a water <u>or</u> energy <u>or</u> mineral ore resource issue in a global or regional setting, for example oil and natural gas production in the Niger Delta, Nigeria

This case study should illustrate and analyse:
+ the themes set out in this unit
+ the relationship between resource security and human welfare
+ attempts to manage the resource.

What do I need to know?	Content and suggested revision methods
How the case study issue relates to the themes set out in this unit (Note that these will depend on your chosen case study)	**For oil and gas production in the Nile Delta** Produce an annotated sketch map that summarises facts relating to energy development and production in the Niger Delta. Annotations should include place-specific facts, e.g.: + In the 1960s, Shell made oil discoveries off the coast near Warri. + In 1977, the state-owned Nigerian National Petroleum Company was set up to regulate the industry. **1 Trade** Make a list of the features of Nigerian oil exports: + Europe 41% + Asia 28% + India 20%. **2 Nigeria's energy mix (2017)** Biomass and waste 74%, oil 16%, natural gas 9%, HEP 1%. Make bulleted notes **to account for** this energy mix, e.g.: + 64% of the population live in rural areas and rely on fuelwood. + As the country develops, demand for oil will increase for industrial and domestic needs.
The relationship between resource security and human welfare	Make revision notes in a table divided into a) resource security and b) human welfare and also into positive and negative implications as this will show **evaluation**.
Attempts to manage the resource	Make a list of bullet point notes to outline how the resource is being managed, e.g.: + Shell paid the Nigerian government for the damage caused by oil spills.

A case study of a specified place to illustrate and analyse how aspects of the physical environment affect the availability and cost of water <u>or</u> energy <u>or</u> mineral ore and the way it is used, for example water resources in Shimla, northern India

What do I need to know?	Content and suggested revision methods
Details of how the physical environment affects water availability	On a piece of A3 paper, create a diagram that incorporates revision notes based on the five sections outlined under 'What do I need to know?'. An example of diagrams showing the first three sections is:

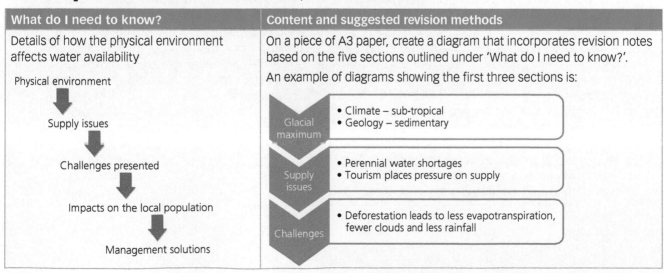

1 Explain the challenges presented by effective nuclear waste management. [6]

2 Account for the patterns shown in Figure 11.15. [6]

Oil
- 25% Others
- 20% USA
- 7% South and Central America
- 9% Middle East
- 5% India
- 13% China
- 3% Russia
- 4% Japan
- 14% EU-28
- Total energy 4,621.9 Mtoe

Coal
- 15% Others
- 9% USA
- 2% Africa
- 11% India
- 6% EU-28
- 3% Japan
- 2% Russia
- 51% China
- Total energy 3,731.5 Mtoe

Natural gas
- 25% Others
- 20% USA
- 5% South and Central America
- 15% Middle East
- 1% India
- 7% China
- 12% Russia
- 3% Japan
- 13% EU-28
- Total energy 3,156 Mtoe

Figure 11.15 Patterns of global energy consumption of fossil fuels, 2017
Source: BP Statistical Review of World Energy

3 Analyse the role of China in the development of mineral ores. [9]

4 Examine the factors leading to changing patterns of energy production and consumption. [9]

5 Referring to a located example, assess the environmental impact of a major water supply scheme. [9]

Answers and quick quizzes online

Opportunities to practise geographical skills within this topic include:

+ use of statistical data, for example charts showing the main producers and consumers of different mineral ores
+ techniques such as frequency distributions, measurements of correlation and testing for the differences between data sets (Student's t-test and Mann-Whitney U test) can be used to interpret and manipulate data
+ map analysis – to determine the global patterns of resource supply

+ use of a range of charts, maps and data sources to present the production, consumption and trade of different energy resources
+ fieldwork skills: data collection, recording, presentation and analysis, for example relating to a water resource issue
+ report writing and the use of core ICT skills based on the collection, analysis and presentation of secondary (and primary where appropriate) data.

Summary

+ Resources can be split into categories of renewable, non-renewable, recyclable, non-recyclable, stock and flow.
+ There is a variety of terms to explain resource evaluation: measured, indicated, inferred and possible.
+ Resource development is a continuous process, which includes addressing sustainability and environmental impact.
+ It is important to have updated and current data on the global pattern of production, consumption and trade of the main energy resources: coal, oil, gas, nuclear and renewable.
+ It is important to have updated and current data on the global pattern of production, consumption and trade of the main mineral resources.
+ Water is unevenly distributed. There are various measures of water scarcity – physical, absolute, economic – and data may also express water stress.

+ Energy, water and mineral resources are all influenced by geopolitical issues.
+ Physical geography influences the supply of energy, water and mineral resources.
+ It is important to be able to evaluate the impact of resource extraction and, in so doing, to refer to specific, detailed examples.
+ Resource consumption must be managed in a sustainable manner to ensure future supplies.
+ All countries aim for resource security, which will reflect a range of economic, political and environmental considerations.
+ Ensuring future supplies of energy, water and mineral resources involves consideration of technological, environmental, economic and political developments.

Glossary

Term	Definition	Page
Adaptation	A change that allows survival or coping mechanisms.	99, 161
Aeolian	Relating to wind action.	40
Arid	An area that receives less than 250 mm of precipitation per year.	36
Aridisols	Soils which form in arid or semi-arid climates.	37
Ash	Dust-sized particles of rock produced by explosive volcanic eruptions.	106
Asthenosphere	The layer of upper mantle extending 100 km to 300 km, slow-flowing and viscous.	101
Biomass	Dry mass of living organisms in a given area or ecosystem at a given time.	127
Biome	A large-scale ecosystem, usually at a continental scale, based on distinct plant and animal species which depend on particular climatic patterns.	122
Capital	Accumulated wealth, which may include machinery, buildings, money, stocks or investments.	143
Carbon pool	Carbon stores.	27
Carbon sequestration	The process of capturing and storing CO_2.	25
Carbon sink	A carbon store that absorbs more carbon than it releases.	26
Conglomerate	A business made up of a number of other, sometimes unrelated, businesses.	145
Containerisation	A system of standardisation that uses large, standard-sized containers for transport.	145
Continentality	The effect on the climate of increased distance from the sea.	39
Decentralisation	The outward movement of people and activities from established centres.	178
Deindustrialisation	The reduction of industrial activity in a region or country.	146, 178
Deregulation	The removal of government rules, regulations and laws from the operation of business.	143
Desertification	Land degradation in dryland areas due to overexploitation by humans and natural processes such as drought.	35
Economic multiplier	The process by which growing economic activity in an area creates employment; employees have money to spend on goods and services and more economic growth is stimulated.	148
Ecosystem	A community of living organisms, their relationship to each other and the environment.	24, 121
Ecosystem services	The benefits people obtain from ecosystems.	124
Edge cities	Self-contained settlements beyond the city boundary.	183
Energy security	The uninterrupted availability of energy sources at an affordable price.	236
Enterprise zone	An area set up by the government to attract industry by the removal of taxes and restrictions to development.	154
Fatalism	The hazard is inevitable and people cannot influence the outcome.	99
Fauna	The animals of a particular region, habitat or geological period.	93
Fetch	The distance of open water over which wind blows without being interrupted by land obstacles.	62
Flood hydrograph	A graph showing river discharge over a period of time when the river's normal flow is affected by a flood event.	17
Flora	The plants of a particular region, habitat or geological period.	93
Flow/transfer	A form of linkage between one store/ component and another that involves movement of energy or mass.	13
Food chain	The sequence of transfer of nutrients and energy from one organism to the next in the order in which they eat one another.	126
Food security	Access to safe and nutritious food in sufficient quantities for individuals to lead a healthy life.	114, 200

My Revision Notes: AQA A-level Geography Second Edition

Term	Definition	Page
Food web	A matrix of feeding relationships that resembles a web.	126
Footloose	An industry that can be placed in any location and is not affected by factors such as resources and transport.	145
Frequency	Distribution over time.	107
Gentrification	The buying and renovating of properties in more run-down areas by wealthier individuals.	182
Geopolitics	The way in which geographical factors shape international politics.	239
Glacier	A large mass of ice moving downhill under the influence of gravity.	75
Global governance	The way in which global affairs are managed through norms, laws, regulations and institutions.	157
Globalisation	The process by which the world is becoming increasingly interconnected.	143
Greenhouse gas	Any gas in the atmosphere that allows short-wave UV solar radiation to pass through into the atmosphere but prevents terrestrial infrared radiation from escaping into space.	24
Hazard perception	The way in which people view the threat of the hazard.	99
High-energy coastline	Has high-energy waves and erosion will exceed deposition.	58
Homogenise	To make a place uniform or similar.	166
Industrialisation	The process by which an economy is transformed from primarily agricultural to one based on the manufacturing of goods.	174
Input	The addition of matter and/or energy into a system.	9
Island arc	A chain of volcanic islands which form during subduction.	103
Lag time	The time between peak rainfall and peak discharge.	17
Land degradation	The deterioration of land.	204
Lava	Hot molten rock from a volcano.	106
Leakage	The economic loss of profits back to companies owned outside of the host country.	143
Low-energy coastline	Has low-energy waves and deposition will exceed erosion.	58
Magma	Molten rock found beneath the Earth's surface.	103
Magnitude	The size of the impact of a hazard event.	107
Marine processes	Processes connected with the sea operating upon a coastline, for example waves and tides.	60
Megacity	A city with a population in excess of 10 million people.	174
Mitigation	The action of reducing the severity of something.	161
Morbidity	The incidence of ill health.	208
Mortality	The number of deaths in a population.	208
Multiple deprivation	The lagging behind in a number of related aspects of life, such as employment, housing and services.	183
Natural hazard	Events that are perceived to threaten people and the built and natural environments.	98
Offshoring	Relocating some part of a firm's activity to another country.	154
OPEC	Organization of the Petroleum Exporting Countries.	240
Outsourcing	The process of subcontracting part of a firm's business to another company in order to save money.	154
Permafrost	Permanently frozen soil and regolith (loose rock overlying bedrock).	203
Plumes of magma	Hot columns of magma rising from deep within the Earth.	104
Prediction	The ability to forecast hazardous events so warnings can be given and action can be taken.	108
Pressure melting point	The temperature at which ice melts under pressure.	80
Primary effects	The effects resulting directly from the event, for example lava and pyroclastic flows.	106

Check your understanding and progress at **www.hoddereducation.co.uk/myrevisionnotesdownloads**

Term	Definition	Page
Protectionist measures	Policies of erecting barriers to trade, for example quotas and tariffs.	145
Regeneration	The revival of urban areas.	179
Resilience	The extent to which an area can recover from the impact of something negative, for example an oil spill or over-fishing.	161
Resource frontier	A newly colonised region where resources have been discovered and are brought into production for the first time.	232
Resource peak	The time of maximum production of a reserve or of a resource as a whole.	232
Sea floor spreading	The theory that the ocean floor is moving away from the mid-Atlantic ridge and crust must be destroyed elsewhere.	102
Secondary effects	The effects resulting from the impact of the hazard, for example flooding and tsunamis.	106
Semi-arid	An area that receives between 250 mm and 500 mm of precipitation per year.	36
Soil organic carbon	The organic components of soil, e.g. tissues from dead plants and animals.	31
Store	A part of the system where energy/mass is stored or transformed.	9
Storm surge	A rapid rise in sea level in which high winds (in a tropical storm) push the sea upwards and in the direction of the coastline.	114
Tariffs	Taxes on imported goods.	145
Tectonic plate	Rigid sections of the Earth's crust that float on the upper mantle and move relative to one another.	102
Tides	The rise and fall in sea level in response to the gravitational pull of the Sun and the Moon.	57
Transnational corporations	Companies operating in at least two countries with a headquarters in one country and other operations (branch plants) usually in a number of others.	154
Undernutrition	Too little food to maintain a healthy body weight.	200
Wave-cut platform	A gently sloping, smooth platform at the base of a cliff, caused by abrasion.	63
Weathering	The breakdown of rocks in situ by a combination of weather, plants and animals.	25

Now test yourself answers

Chapter 1

1 In an open system there is a transfer of energy and matter, in a closed system just energy.

2 a) Water is added to the hydrosphere from weather elements in the atmosphere, such as rain and snow. Water from the hydrosphere evaporates back into the atmosphere.

   b) Soils form from broken-down rock (lithosphere) and supports plants and animals in the biosphere.

   c) Frozen water in the form of glaciers (cryosphere) erodes and breaks down the rocks in the lithosphere.

3 As populations grow, the demand for water increases for domestic, agricultural and industrial needs. Underground water stores are increasingly sought to fulfil water supply, especially in agricultural areas and in poorer countries where deep wells can be sunk.

4 Low rainfall → low soil moisture → reduced transpiration → less rainfall.

5 Permanently frozen soil and regolith (broken-down rock fragments).

6 Ant three of: rivers, lakes, wetlands, groundwater, soil water, biological water.

7 Water vapour absorbs and reflects incoming solar radiation. Increases in water vapour lead to increases in atmospheric temperatures.

8 a) Water evaporates from the surface of the Earth and condenses around nuclei to form visible water droplets (clouds).

   b) Water evaporates into the atmosphere; condensation occurs (when air temperature reaches its dew point or due to adiabatic cooling) forming rainfall.

9 a) Warmer temperatures lead to higher rates of evapotranspiration, as warm air can hold more water vapour.

   b) Evapotranspiration increases as wind moves humid air away and the air does not become saturated as quickly.

   c) The more humid it is, the lower the evapotrans-piration as the air becomes saturated quickly.

10 On sunny days, the air is heated by warm surfaces, it rises rapidly, cools, condenses and forms convectional rainfall. The rainfall is a short, heavy burst due to the rapid processes of heating and uplift.

11 When climate change results in an increase in dry, hot weather and aridity, soil moisture stores decline.

12 Precipitation (input), vegetation density and ground water saturation can affect storage in the drainage basin system.

13 Saturated soil, due to rainfall exceeding soil infiltration capacity, or impermeable surfaces (for example concrete) leading to an increase in overland flow.

14 Ocean evaporation adds water vapour to the atmosphere as water turns to vapour. Most of this falls back into the oceans as rain; some falls over land surfaces as rain. Increased ocean evaporation and therefore water vapour adds water to the water cycle. It can also increase precipitation in storm events.

15 In autumn and winter there will be more rainfall, ground will become more saturated and groundwater stores will increase if the ground is saturated. Overland flow will also increase. In the summer months, with higher temperatures and less rainfall, groundwater stores will reduce.

16 Water balance is the long-term balance between inputs and outputs in the drainage basin system. It is a balance between precipitation, evapotranspiration and river discharge.

17 Climate: temperature, precipitation. Drainage basin characteristics: gradient, geology, drainage basin size, shape, density, land use.

18 Base flow is water input to streams and rivers, it represents the normal day-to-day river discharge level.

19 Shorter lag time: saturated ground, high levels of rainfall, high discharge level/above normal already.

   Lengthening lag time: low levels of groundwater, high temperatures, low level of discharge due to dry period, small amount of rainfall/light rainfall in the storm.

20 The Pacific and Atlantic oceans (the latter in particular) circulate water from the ocean to landmasses. The climate over the tropics means that there are high levels of evaporation due to warm temperatures and water vapour turns to rainfall.

21 Biosphere: in organic matter in soils, plant litter, soil humus, organic molecules in living and dead organisms.

   Hydrosphere: in plankton that photosynthesises, decomposing plant and animal matter, living and dissolved organic matter, calcium carbonate shells in marine organisms.

22 Sedimentary rocks.

23 The rate of energy transfer per unit area, for example respiration, photosynthesis, diffusion, combustion.

24 Photosynthesis.

25 Organic matter, which can be vegetation or fossil fuel.

26 Fast carbon cycle: terrestrial, relates to the uptake of carbon in the process of photosynthesis and the release of carbon dioxide to the atmosphere in

respiration and decomposition; the cycling of carbon between soil, vegetation and atmosphere is rapid.

Slow carbon cycle: the cycling of carbon between rocks, the atmosphere and the oceans. It is done through the process of weathering, which takes millions of years.

27 Mainly by photosynthesis and water absorption.

28 The capture of carbon from the atmosphere or from anthropogenic sources, for example power stations.

29 Decomposition is faster in warm climates as there is more bacterial activity than in cold climates.

30 A carbon flux is the rate of energy transfer in the carbon cycle; a carbon store is the quantity of carbon in a carbon 'pool'/store.

31 When soil is eroded it is transported and eventually deposited. The deposition phase can create a carbon pool where the carbon will be stored for many years, for example when particles are deposited in the sea.

32 Urban growth reduces the amount of surface vegetation, which impacts the carbon cycle. Replanting projects to compensate for urban development and the inclusion of open spaces with vegetation in the planning of new urban areas would both help offset the damaging impact.

33 The net removal of carbon dioxide from the atmosphere by plants and micro-organisms and its storage in vegetation biomass and in soils.

34 The increased plant growth resulting from an increase in atmospheric carbon dioxide. Increased carbon dioxide is converted to plant matter through the process of photosynthesis.

35 The impact on climate of the additional heat retained in the atmosphere due to increased carbon dioxide and other greenhouse gases resulting from human activity.

36 Albedo is the proportion of sunlight reflected from surfaces. Bright surfaces reflect sunlight, leading to warmer climates, snow melt, less reflected sunlight, warming and a warmer climate.

37 There is uncertainty about the full effect of clouds on climatic feedback because there are different types of cloud. High clouds retain heat for longer, whereas low clouds are thicker and reflect more sunlight.

38 Livestock, decay of organic waste, the deep oceans.

Chapter 2

1 Sun's energy, precipitation, wind.

2 Their boundaries are open to both inputs and outputs of energy and matter.

3 Positive feedback: progressively greater change from the original condition. Negative feedback: the system is returned to its original conditions.

4 A system in which constant changing is occurring to achieve balance. Balance is maintained by adjustments to inputs and outputs.

5 A wide daily temperature range – for example, temperatures of over 30°C during the day and lows

of 0°C at night. The heat escapes rapidly at night as there is little or no cloud cover.

6 Convectional rainfall – rapid heating of land surfaces leads to uplift of rising hot air, which cools, condenses and falls as intense bursts of rainfall.

7 a) Due to slow rates of weathering and a lack of vegetation

   b) as evaporating moisture leaves salt behind.

8 Any three of: they are able to store water in the cuticle; they have long root systems to reach deep water supplies; they have short roots to catch brief spells of precipitation; they are drought evading (they germinate and set seeds when it rains, the seeds remain dormant until the next rains); they are perennials, which lie dormant during dry spells and come to life when water is available; they can survive in saline conditions; they have small leaves, stomata that close during the day and waxy cuticles, all to reduce water loss.

9 The ratio of precipitation and potential evapotranspiration (a numerical indication of dryness of the climate in a given location).

10 Cells of circulating air in the tropics. Rising warm air at the Equator is replaced by cooler air drawn in from the north and south, creating areas of low pressure – the intertropical convergence zones (ITCZ).

11 As there is no cloud cover, land heats quickly during the day to a high temperature. As a result of few clouds, heat escapes quickly at night, leading to a large diurnal temperature range.

12 An area of below-average precipitation situated in the lee of an upland area.

13 Incoming solar radiation, received either in the Earth's atmosphere or at its surface.

14 ✤ **Thermal fracture** (expansion and contraction of rock).
   ✤ **Exfoliation** (salts brought to the surface by capillary action in plants lead to chemical weathering).
   ✤ **Crystal growth** (pressure and expansion as salt crystals form and grow).
   ✤ **Hydration** (absorption of small amounts of water leading to further chemical weathering).
   ✤ Block and granular disintegration from **freeze–thaw weathering**.

15 Change in temperature – heating and cooling, the presence of even a small amount of water, living organisms.

16 Block: breakdown of rock into large blocks resulting from joints and bedding planes being weathered. Granular: much finer, individual grains are weathered away from rock surfaces.

17 Physical weathering often involving weather elements, for example precipitation and temperature change.

18 Aeolian (wind) deflation, corrasion and attrition. Fluvial – river erosion – abrasion, attrition.

19 Wind erosion is the abrasive effect of sand grains close to the desert surface; deflation is the removal of sand by wind. Wind transportation is the

My Revision Notes: AQA A-level Geography Second Edition

saltation and creep of sand grains and fine sand and clay particles being carried in the atmosphere.

20 Intense overland flow of water, often as the result of an intense burst of rain running over hard-baked land surfaces.

21 Dew, rainfall – often short-lived heavy downpours, perennial rivers, seasonal river catchments.

22 A short-lived stream that flows after a heavy downpour.

23 Ventifacts are small rocks that have been abraded or shaped by wind-blown sand.

24 Barchans form when wind comes predominantly from one direction. The windward slope is much gentler than the leeward slope. The wind will be strong and frequent, providing an adequate amount of sand to form a dune.

25 A star dune is formed where the wind comes from different directions creating a pyramidal profile with slip faces on three or more sides.

26 Deflation, abrasion and attrition.

27 The degradation of formerly productive land to the point where desert-like conditions prevail.

28 Population growth results in an increased demand for food and an increased need for wood for fuel. This leads to more land being farmed and in a more intensive way, and to deforestation. Vegetation is removed; there is more soil erosion, more evaporation of moisture, and over time, desertification occurs.

29 Reduced habitats due to reduced vegetation. Less nutrient cycling in soils. Soil is lost due to dryness and exposure. Loss of biodiversity, food chains and webs become more fragile.

30 When the population increases, people do not have money to buy more food as they are based in a subsistence agricultural system. Also, they cannot afford the resources required to manage the problems resulting from desertification.

31 Small-scale approaches that are planned and controlled by local communities. They often have a low environmental impact.

32 Technology that is 'appropriate' in its context of use, i.e. small-scale technology, it can be managed locally, and it often uses skills available in the local community.

33 Any two of: shelterbelts can be planted to reduce soil erosion and increase infiltration of rainfall; reduce population growth; global attempts to reduce climate change (for example less use of fossil fuels); invest in local knowledge to manage soil conservation and water supplies.

34 Due to low biodiversity; highly specialised and adapted plant and animal species with sparse populations; slow plant growth due to aridity; short food chains; the loss of one species can have a huge impact; high rates of soil erosion when vegetation is removed.

Chapter 3

1 They allow the transfer of both energy and matter. They have inputs, stores and outputs.

2 When balance is maintained by adjustments to inputs and outputs; the system is constantly changing.

3 From waves, tides, wind and sea currents.

4 Geology will determine how effective the energy inputs can be in the weathering and erosion of rock and over time in the landforms produced. A coastline combining soft rock and a high energy input will lead to a landscape of certain characteristic landforms (for example bays and beaches), whereas a geology comprising more resistant rock will yield a different landscape depending on the energy input.

5 Wind that is most dominant or most usual. Wind energy input will be concentrated from the prevailing wind direction and this will affect geomorphological processes of erosion and deposition.

6 The length of water over which a given wind has blown without obstruction (for example from land masses).

7 The greater the wind speed, the longer the duration of the wind, and the longer the length of fetch, the greater the wave energy and the larger the wave. Slow wind speed and less wind results in smaller waves.

8 Wave refraction is the process by which waves break onto an irregular coastline. The waves drag in the shallow water approaching a shoreline, the wave becomes steep and short, and the part of the wave in the deeper water moves faster. The wave bends. The low-energy wave spills into the bay and most of the energy is concentrated on the headland.

9 An ocean current is a large-scale current generated by the Earth's rotation and by convection. Ocean waves are generated by the wind at the ocean surface; they can travel long distances to the shore from open waters, moving in a circular motion.

10 Rip currents and longshore drift.

11 As waves enter shallow water, the circling water molecules come into contact with the seabed. Friction changes the speed and direction of the wave. The wave slows due to frictional drag, wavelength decreases and waves start to 'bunch up'. The bottom of the wave slows, the top of the wave steepens, advances and eventually topples over and breaks on the shore.

12 A stretch of coastline within which the movement of sand and shingle is largely self-contained.

13 Seabed sources from continental shelf areas. River deposits (90 per cent). Sediment from cliff erosion (5 per cent). Shells and corals.

14 Interventions to manage the coast, such as beach nourishment where sediment is added.

15 Wind direction, tidal currents.

16 ✚ A long length of fetch means that strong waves build up with a high level of erosional power.
    ✚ Constructive waves (low height, long wavelength) and low-frequency waves – sediment movement off the beach is low. Destructive waves (greater height, short

Check your understanding and progress at www.hoddereducation.co.uk/myrevisionnotesdownloads

wavelength, high frequency) have more of an impact on erosion.

+ Beach material will protect the coastline and therefore coastal erosion will be less.

17 Weathering is the in-situ breakdown of rocks exposed at or near the surface. Mass movement is the downslope transportation of material under gravity.

Mass movement processes will add material to the sediment budget (the balance of the sediment volume entering and exiting a particular section of coast). The exact impact of this will depend on a wider set of processes and contextual factors for any particular stretch of coastline.

18 Hydraulic action is the force of the water on the rocks. Wave quarrying is when a breaking wave traps air in cracks in a cliff face; when the water pulls back, air is released under pressure and this weakens the rock face over time.

19 Three from: oxidation, hydration, hydrolysis, carbonation.

20 Climate determines the amount of moisture available, moisture on a slope leads to an increase in mass movement. Freeze–thaw cycles are an example of how climate can affect moisture on slopes.

21 Headlands attract energy due to wave refraction.

22 Concordant coastline: a coastline of bands of different rock type lying parallel to the shore. Discordant coastline: bands of different rock type lie perpendicular to the coastline.

23 Geology (hard or soft rock); climate (affects weathering processes); human activity (for example beach nourishment).

24 Weathering processes (mechanical, chemical and biological) weaken the rock, erosion processes attack the cliff face and remove and use weathered material; marine erosion processes (including hydraulic action, wave quarrying, abrasion, attrition, solution) attack the cliff face and remove and use weathered material.

25 The coastline will have characteristic features of cliffs, wave-cut platforms, caves, arches, stacks. There will be rugged cliff faces and no long, extended beaches.

26 The movement of sediment by waves and currents along a shoreline. It will follow the direction of the prevailing wind, with the swash washing sediment on shore and the backwash taking it back down the beach.

27 A swash-aligned beach will develop where there is minimal longshore drift and the coastline is irregular. A drift-aligned beach forms parallel to the direction of longshore drift.

28 Material carried by longshore drift will continue to be deposited across a bay or where the coastline changes direction, forming a spit. If a spit develops across a bay and there is no strong flow of water from the landward side, a bar is formed. A spit that joins an island to the mainland is a tombolo.

29 Saltation (bounding action of wind-carried sand particles) and creep (the surface movement of larger sand particles).

30 Vegetation forms a barrier, which reduces the impact of wind at the surface and traps sand. Marram grass is specially adapted to survive in a dune environment and can grow tall and dense, trapping sand.

31 a) A waterlogged area between two sand ridges.

b) A rapid erosion of sand dunes where vegetation has been removed, often caused by human activity.

32 A global change in sea level resulting from a rise or fall in the level of the sea itself (for example as a result of the retreat of ice following a glacial period).

33 Where an expanse of gently sloping, formerly submerged land has been exposed by uplift or the lowering of sea level.

34 A drowned glacial trough (U-shaped valley).

35 Coastal flooding, more coastal erosion (as waves attack areas previously above the high-tide level), receding coastlines, more storm surges, increased need for and investment in coastal protection measures.

36 Investment in coastal protection, increasing insurance costs and the cost of insurance claims if land is lost to coastal erosion.

37 Longshore drift.

Chapter 4

1 Solar radiation, geothermal heat, mass from rockfall and snow (direct snowfall, windblown or avalanches).

2 The balance between inputs and outputs. In a glacier, the balance is dynamic (constantly changing), with the glacier constantly gaining inputs and losing outputs. If the inputs and outputs are the same, the glacier is in dynamic equilibrium.

3 The difference between annual accumulation and annual ablation – it may be positive, negative or neutral.

4 + Polar is below freezing all year; tundra: has a short summer thaw of temperatures of +5°C.

+ Polar has a winter average of below –50°C and tundra has a winter average of –20°C.

+ Polar has less precipitation (150 mm per year) and tundra has averages of 300 mm per year.

5 Input received in the form of short-wave solar energy.

6 The proportion of sunlight reflected from a surface.

7 Due to strong westerly winds, cold oceans and very large land masses.

8 Generally over 3000 m in elevation; landscapes include ice caps, mountain glaciers and tundra; landscapes develop over glacial and interglacial periods.

9 The variation in temperature between –10°C in winter and 20°C in summer creates a climate for both ice- and water-based erosional processes.

10 Largely treeless environments with short growing seasons and severe winter temperatures leading to areas of permafrost (permanently frozen ground).

11 Due to the grey ferrous iron compound content of the soil.

12 Three from:
+ Low productivity and growth rates.
+ Perennials which can store water from year to year.
+ Close to the ground for protection from strong winds.
+ Shallow root systems to capture water from a brief summer thaw.
+ Can photosynthesise at very low temperatures.

13 The boundary between snow-covered areas and areas with a higher temperature and no snow cover.

14 The boundary between the accumulation (input greater than output) and ablation (output greater than input) areas of a glacier.

15 The difference between accumulation and ablation in a year.

16 Glacial advance will result during a period of accumulation and glacial retreat in a time of ablation.

17 A layer of compact snowflakes.

18 Warm-based glaciers occur in temperate areas. They are small, with summer melts. This lubrication means more movement and more erosion. Cold-based glaciers are often frozen to the bedrock, meaning less erosion.

19 Physical (mechanical) weathering: involves processes that break down rocks without altering their chemical composition. Often these processes involve elements such as water or ice.
Chemical weathering: involves chemical reactions in processes such as oxidation, solution and hydrolysis.

20 a) Ice moves with brittle, breaking movements.
b) There is more meltwater and therefore movement is 'sliding' and more rapid.

21 Extensional flow occurs over a steeper gradient and the movement of ice accelerates. Compressional flow is where there is a reduced gradient, the ice mass thickens and movement is slower.

22 + On the surface (supraglacial)
+ within the ice (englacial)
+ at the base of the glacier (subglacial).

23 It is angular in shape, unsorted and unstratified, and dropped in mounds rather than layers.

24 Surface water lying over the frozen layer melts but cannot drain away. It therefore lubricates soil particles which move downslope.

25 Freeze–thaw predominantly, some plucking.

26 a) A deep crevasse that opens up between the glacier and the back wall of an ice-filled hollow.
b) Truncated spurs are formed when areas of land protruding from the valley side (spurs) are removed by the glacier.

27 Roches moutonnées form on the valley floor as areas of more resistant rock are not completely removed by ice movement. They have a smooth up-valley side created by glacial abrasion and a more jagged down-valley side created by plucking.

28 Through a combination of erosional processes: abrasion (debris embedded in the glacier 'scours' the surface of rocks, wearing them away) and plucking (meltwater seeps into joints and cracks, freezes, becomes attached to the glacier, the ice moves and pulls fragments of rock away with it).

29 Changes in the rate of flow of a glacier or the presence of areas of soft rock on the valley floor mean that some parts of the valley floor may be deepened and form a long, thin hollow which is filled with water when the ice retreats, creating a lake called a ribbon lake.

30 From the reshaping of previously deposited glacial material, the accumulation of material around a bedrock obstruction, or the thinning of ice in a lowland area leading to deposits of debris.

31 Recessional moraine occurs where there is an interruption in the retreat of the ice. Push moraine forms if the climate cools and the ice advances. Recessional moraine is associated with ice retreat and push moraine with ice advance.

32 Due to high levels of discharge; energy declines with a reduced discharge.

33 Braided channels are water channels that have divided. They are common in areas of deposition as the large amounts of deposited debris 'choke' the channels and cause them to split.

34 When blocks of ice washed onto the outwash plain melt and leave a gap in the sediments.

35 Periglacial areas are not glaciated but are exposed to very cold conditions. They include tundra landscapes.

36 They include long periods of sub-zero temperatures and a short summer melt when meltwater lies over areas of permafrost.

37 The expansion of soil volume as ice crystals form within the soil.

38 Closed system pingos: generally found in areas of permafrost; develop beneath lake beds; the growth of the ice core is hydrostatic. Open system pingos: generally found in areas of discontinuous permafrost; found in valley bottoms; the growth of the ice core is hydraulic.

39 The slow flow of fine, water-saturated soil (following a brief period of melting) and fine rock fragments from high to low ground, over the smallest of gradients.

40 There have been eight glacials in the last 740,000 years. Each glacial advance has altered the landscape left by previous glacial periods. The pace of the processes that shape the land varies over time.

41 Due to the low levels of biodiversity resulting from the harsh conditions and the delicate thermal balance. Food chains are short, meaning that disruption at any level can have long-term and far-reaching impacts. This leads to a slow and uncertain recovery.

42 It is managed by several laws and treaties including the Antarctic Treaty (1959), the Madrid Protocol (1998) and conventions – Convention for the

Conservation of Antarctic Seals (1972), Convention for the Conservation of Antarctic Marine Living Resources (1980) and Convention on the Regulation of Antarctic Mineral Resources (1988).

43 Pressure to develop the fuel and mineral resources (for example coal, oil, precious metals); pressure to develop economic activities such as tourism; pressure to expand fishing rights, as stocks elsewhere become depleted; pressure to exploit the biochemical resources of the flora and fauna.

Chapter 5

1 A water-related hazard, for example floods, but also mudslides and landslides.

2 Reducing the severity of the impact.

3 Its magnitude and duration. It is also determined by a range of environmental, social and economic factors, such as mitigation, experience, perception, preparation, physical setting, technology for warnings and responses, wealth and vulnerability.

4 Socio-economic status, education, employment, culture, past experience and values.

5 The sustained effort of communities to respond to and withstand the impact and effects of hazard events.

6 A combination of how a hazard is managed pre-event (for example in terms of prevention and preparation), how the event itself is responded to (for example, in terms of warnings, evacuation, assistance and restoration) and how post-disaster planning and management are approached (for example in terms of recovery and reconstruction).

7 Natural hazards exist in the interface between the physical and human environments. A hazard becomes a disaster when there is loss of life and/or destruction of the built environment and/or disruption of human activities.

8 Plates converge at destructive plate margins. There are three types of destructive plate margins. At oceanic and continental plate margins, typically there is subduction, formation of ocean trenches, earthquakes and volcanic activity. At oceanic-to-oceanic destructive plate margins, typically there is subduction and volcanic activity. Finally, at continental-to-continental destructive plate margins, there is little subduction and formation of mountains due to uplift.

Where plates diverge at constructive plate margins, features such as rift valleys and ocean ridges form, and energy released from plate movement causes earthquakes.

9 Scientists now believe that plates move as part of a gravity-driven system which is the combination of two concepts – ridge push and slab pull.

10 When two oceanic plates converge. During subduction, as the more dense plate dips below the less dense plate, the descending plate encounters hotter surroundings which together with friction cause the plate to melt. Less dense material rises to the surface and forms explosive volcanoes which can form a line known as island arcs.

11 Where two low-density continental plates converge there is not much subduction due to meeting of similar, low-density material. Sediments between the two plates are forced up to form young fold mountains. The Himalayas are formed where the Indo-Australian Plate (continental plate) is being forced northward into the Eurasian Plate (continental plate).

12 Ocean ridges form where plates move apart in oceanic areas, rift valleys are formed where plates move apart in continental areas.

13 An area of intense volcanic activity where a mantle plume reaches the Earth's surface, causing eruptions.

14 Oceanic crust is relatively thin (5 km deep on average) and composed of dense basalt rock. Continental crust is mainly granite, which is less dense and is on average 30 km deep or can be up to 100 km deep beneath mountain ranges.

15 When plates move apart in continental areas. The Earth's crust fractures as sections of it move. Areas of crust drop down below parallel faults, forming a wide, deep rift valley.

16 At constructive plate margins magma flows to the surface under reduced pressure. Lava, tephra and hot gases are part of the eruption. Lava is basaltic – it has low viscosity and eruptions are less violent. Gases escape easily from basalt. At destructive plate margins eruptions are more violent and composed of viscous, thick, andesitic lava and tephra.

17 A combination of gas and tephra, which is extremely hot (over 800°C) and which flows down the side of a volcano at speed.

18 Two from:

+ Volcanic mud flows: lahars – a combination of melted snow and ice, rock, sand and volcanic ash.
+ Floods: from melted glaciers and ice caps.
+ Acid rain: gases from volcanoes include sulfur this combines with atmospheric moisture to form acid rain.
+ Tsunamis: the displacement of ocean water creating a huge wave which increases in height on approaching the shore.
+ Mudflows and landslides: mass movements of material which becomes lubricated with flood water.

19 The viscosity of the lava (that is more viscous lava leads to slower flow), steepness of the ground, and whether the lava flows as a broad sheet or as a lava tube.

20 Mitigation attempts include prediction by the study of land swelling, groundwater levels and chemical composition; gas emissions at the surface; detection of shock waves.

Protection measures include evacuation; hazard drills; land-use planning; controlled explosions to divert lava flows.

21 The process by which sediments and soils lose their mechanical strength from a sudden loss of cohesion. The material is transformed temporarily into fluid.

22 At destructive plate boundaries, plates of mostly oceanic crust are sinking (or being subducted) beneath another plate. This causes earthquakes in the subduction zones due to the plates slipping past one another and rupturing within the plates.

23 This can be due to the reactivation of old fault lines.

24 Primary hazards: ground shaking and splitting. Secondary hazards: shockwaves, tsunamis, liquefaction and landslides.

25 Size of event: the more powerful, the greater the impact. Population density: a high population density leads to a greater impact on people and their property. Degree of preparation: the more educated, drilled and prepared people are, the lower the impact. Time of day: at night there is more risk of people being caught unaware and therefore not having time to protect themselves. Level of economic development: higher levels of development mean there are likely to be more effective mitigation strategies in place.

26 Prediction: including release of radon gas, monitoring groundwater levels and animal behaviours, measuring magnetic fields and studying hazard zone maps.

Protection: including developing understanding, safety drills, building modifications (rubber shock absorbers, cross-bracing), education, land-use planning, tsunami protection and warning systems.

27 Ocean depth of more than 70 m, a location 5°N or 5°S of the Equator, convergence of air in the lower atmosphere, rapid outflow of air in the upper atmosphere.

28 Warm ocean water causes a large amount of water evaporation; winds converge close to the ocean surface; air is forced upwards; the air is unstable and winds rise rapidly; warm rising air cools and condenses to form cloud and rain; the heat generated from the rapid condensation warms the surrounding air; it rises, forming an intense 'up draught'; dry, cooler air from the upper atmosphere descends.

29 Onset is rapid and their paths are erratic.

30 Hurricane and cyclone drills, rapid evacuation, land-use planning, sea walls, breakwaters, flood barriers, damage reduction through building design that makes buildings and structures such as bridges more resistant.

31 Three from: storm intensity, speed, distance from the sea, preparation, warnings.

32 Flooding, landslides, mudslides.

33 If controlled, small fires can prevent the build-up of fuel, which might contribute to large, more destructive and dangerous fires. The ashes add nutrients to the soil. They can provide a means to control pests and alien plant species.

34 Damage to power lines, arson, careless discarding of cigarettes and lack of attention to campfires.

35 Low cost: land-use planning and education. High cost: aeroplanes to spray water, extensive emergency response cover.

36 Without insurance, people can lose their homes with no help to rebuild. Toxic pollutants can stay in the atmosphere for long periods of time and have widespread effects. Damage to ecosystems and wildlife can be long term or may be irreversible if a species is wiped out.

Chapter 6

1 Biodiversity is the number, variety and variability of living organisms. It includes diversity within and between species. It is part of the global ecosystem and covers how diversity changes from one location to another.

2 Certain species that are used as an indicator of environmental conditions.

3 Three from:
  + Increased global temperatures – alter food supply, breeding patterns, migration patterns.
  + Increased methane levels – add to global warming – leading to habitat and food supply changes.
  + Change in the chemical nature of rivers and estuaries due to urban and agricultural runoff.
  + Grasslands, forests and wetlands have been transformed by humans – for urban growth, increased food production, changing patterns of land use, for example removal of hedgerows in farming.

All lead to habitat destruction and/or degradation and food supply challenges.

4 Organisms are transported across geographical barriers that kept the biological regions of the Earth separated; there is a transfer of non-native species and sometimes these take over from native species or can bring disease problems – both can reduce biodiversity.

5 Where populations depend on indigenous plants for food supply, declining biodiversity can impact on these varieties, resulting in them being wiped out or significantly reduced. Also, where insect populations are affected, there can be issues in food supply as insects are a natural pest control.

6 The quality and availability of natural water supply is reduced. Trees intercept rainfall and allow it to percolate through the soil and into groundwater supplies. With fewer trees there is more runoff and a more rapid transfer of water over the land rather than into groundwater supplies.

7 Population growth increases the pressure on ecosystem services and consumption per capita increases. There is more demand for basic supplies of food and water, more damage to ecosystems and, due to human impact, more degradation of soils and natural water supplies.

8 Biotic factors refer to living organisms. They are subdivided into producers, consumers and

decomposers. For example, green plants (producer), herbivores (primary consumer), carnivores (secondary and tertiary consumers), bacteria, fungi (decomposers).

Abiotic factors are non-living things, for example climate, soil, topography, altitude.

9 Two from: photosynthesis, decomposition, feeding/digestion.

10 The flow of energy in an ecosystem occurs in a number of stages called trophic levels. They relate to the different consumers – primary, secondary, tertiary and quaternary consumers.

11 Because at each trophic level energy is lost as organisms convert only a small percentage of the energy they consume into living tissue.

12 Energy from the Sun is absorbed by green plants and used in the process of photosynthesis; this creates carbohydrates. The plants are eaten by primary consumers and energy passes up the food chain through different trophic levels. When organisms die, decomposers return nutrients to the soil to be used by growing plants. Biomass declines at each ascending trophic level as organisms convert only a small percentage of the energy they consume into living tissue.

13 Producers are life supporting as they convert the input of the Sun's energy into usable carbohydrates.

14 Gross primary productivity is the total energy fixed by plants in a community. By subtracting energy used for respiration, net primary productivity is achieved.

15 A vegetation stage as an ecosystem moves through a process of succession (a sequence of vegetation changes through time). Each sere modifies the environment, allowing new species to grow.

16 When the vegetation is in balance with the natural environment. Arresting factors include change in topography (for example a landslide), changes in drainage (for example water table lowers), introduction of an alien species, environmental change through human activity.

17 Environmental factors, such as climate, geology, soil type, microclimate variations, availability of food and shelter.

18 Secondary succession takes place on sites that have been vegetated but where the vegetation has been destroyed (for example by fire).

19 They can either survive the environmental change by forming new adaptations (for example new breeding patterns due to change in climate) or an alien species will outcompete the existing species so they reduce in numbers – eventually leading to extinction.

20 Dead organic matter on top of the soil.

21 Biomass, litter, soil.

22 When soluble materials drain away in soil.

23 Due to the warm, wet climate, plants grow quickly, the recycling of nutrients is rapid and as a result, the storage of nutrients in the soil is very low as they are needed for new growth.

24 Changes in seasonal timings can disrupt the natural life cycle. Food supply for raising newborn animals may also be disrupted.

25 Soils dry out and can be easily eroded; this will impact on plant growth, which in turn will affect the numbers able to survive at each trophic level – fewer primary consumers, secondary consumers, etc. The ecosystem will become fragile and animal populations will be vulnerable.

26 A large-scale ecosystem, usually at a continental scale, which has distinctive plant and animal species.

27 An area largely in the sub-Arctic and high mountains that has a short growing season and severe winters.

28 A hydrosere refers to the process of succession starting in fresh water and a lithosere is where succession starts on an exposed rock surface.

29 Because it takes a long time to break down rock surfaces. The rate of weathering and breakdown will be determined by the climate.

30 So that they can maximise their exposure to sunlight.

31 Between 30°N and 30°S of the Equator where surface water temperatures do not drop below 16°C.

32 If water temperatures in areas of coral reef become too high, the algae that give the corals their colour leave the polyp, exposing the white calcium carbonate skeletons of the coral.

33 Nutrient enrichment in rivers and streams leads to a fall in oxygen levels due to the increased activity of aerobic bacteria, resulting in the subsequent death of species that are dependent on oxygen.

34 Ponds form habitats for a range of important insect species, which in turn have a role to play in the growth and survival of different plant and animal species.

Chapter 7

1 A process that integrates people across the world. It has a range of economic, social, cultural, technological, environmental and political dimensions.

2 Investment by a company into the structures, equipment or organisations of a foreign country.

3 The USA, EU and China are the main recipients of FDI while Japan is the highest investor in FDI. Aid flows from high- to low-income countries. Remittance payments flow mainly from developed regions such as North America and Europe to southern and Southeast Asia.

4 The main advantages centre on the provision of humanitarian relief, for example food supplies, medicine and shelter. The main disadvantage is that it can create dependency rather than self-reliance.

5 Remittance payments can generate economic growth in the recipient country as families have money to spend.

My Revision Notes: AQA A-level Geography Second Edition

6 a) The economic loss of profits back to the companies owned outside of the host country.

   b) The movement of money from migrants working outside of their home country back to their families.

   c) An industry that can be placed in any location and is not affected by factors such as resources and transport.

   d) A system of standardisation that uses large, standard-sized containers for transportation.

   e) The reduction of industrial activity in a region or a country.

7 High-level services are services to businesses, for example finance, investment and advertising. Low-level services are services to consumers, for example banking, travel and tourism.

8 By offering incentives, such as tax breaks, for TNC investment; by upgrading skills and technology; through protectionist policies, for example import tariffs.

9 Economic growth means that wages and employment levels are rising, so people have more disposable income and more spending power, and therefore consumption increases.

10 International trading is now faster and easier than ever before. The global financial system provides a framework to facilitate flows of capital. Electronic trading systems mean that companies around the world can trade rapidly and securely, and deregulation of financial markets has led to the removal of barriers to the movement of finance.

11 They have increased the efficiency of the flow of goods.

12 Where various stages of the production process are located in different parts of the world.

13 Advantages include: economic development, representation in world affairs, freedom of movement of goods and labour, sharing of technological advances.

   Disadvantages include: lack of access to trading blocs forming a development gap, trade disputes arising over tariffs, some loss of sovereignty, pressure to adopt central legislation.

14 More nations are participating in trade than ever before, leading to growing economic interdependence. Countries rely on other countries to supply the goods and services they need and so become more interdependent.

15 Air pollution travels across international boundaries and so a pollution source may be in one country but the effects are felt in other countries.

16 They can mean that a country is too reliant on capital flows from other countries and this does not foster economic self-reliance.

17 They increase the workforce, pay taxes, spend money, promote growth and can reduce dependency in high-income countries with an ageing population.

18 They can access technological, educational and training advancements which in turn can lead to higher-paid jobs and greater prosperity.

19 These countries have less political influence and so growth can be restrained as they are excluded from trading agreements and access to some markets. They have limited power and become dependent on the decisions made by wealthy countries.

These decisions are often in the best interests of the wealthier nations.

They may become overly reliant on aid and other forms of financial support, meaning that independent growth is difficult.

20 They are impacted by decisions and events outside of their control, for example economic recession or change in consumer preferences and demand.

21 Advantage – they can pool resources to help poorer countries and to bring about change on global issues such as climate change.

Disadvantage – the countries included can have a more dominant influence on economic and political global systems.

22 Countries specialise in providing goods and services that they excel at.

23 This happens when a country has a more protectionist policy towards trade and enters into fewer trading agreements.

24 As their economies are not as developed, they rely on foreign investment to create jobs and thereby economic growth.

25 Trade increases employment. Through the upward multiplier effect this leads to economic growth as there is more affluence, more spending and more investment from the government. Trade leads to infrastructure development, which leads to further economic growth. As countries develop and their investment in business, education and training increases, further investment takes place from within the country and beyond, furthering overall economic development.

26 a) Conflict can occur as all countries in a trading situation want to secure the best deal for their citizens and businesses. Conflict can arise when countries are accused of promoting 'protectionist policies', for example the USA holding back on some regional trade agreements.

   b) Trade can be used as a bargaining tool in geopolitical issues. Economic trade sanctions can be used to inflict financial hardship.

27 Price fluctuations are a common feature of primary goods as supply and demand fluctuates. Some primary goods, such as food products, are vulnerable to the impact of climate which can either increase or decrease supply – both having price implications. Manufacturing goods adds value to exports, yielding more income.

28 Over-dependency on a limited range of exports leaves a country open to risks if prices change or supply is affected. Diversification helps to spread the risk.

29 With multiple locations, they can avoid trade tariffs, find low-cost locations for production, reach foreign markets and exploit natural resources.

Check your understanding and progress at www.hoddereducation.co.uk/myrevisionnotesdownloads

30 Outsourcing is a cost-saving strategy used by companies whereby they arrange for goods and services to be produced by other companies at a location where costs are lower. Offshoring is the practice of basing some of a company's processes or services overseas, so as to take advantage of lower costs.

31 The management of global affairs.

32 a) The WHO directs and coordinates international health issues within the UN.

   b) The UN addresses global, economic, social and environmental issues.

33 The global commons are for the use and benefit of *all* people and should not be contested and developed by individual nations. Some countries will have more capital and political power to lay claim to the global commons and this should be protected against.

34 It is almost entirely covered by ice. The continent is divided into the east and west Antarctic ice sheets with mountains in both the east and west of the continent. The coastline has ice shelves – the Ross Ice Shelf and the Ronne Ice Shelf. Nunataks are high mountain peaks that protrude above the ice sheet. The underlying geology is mainly igneous and metamorphic rock. West Antarctica is part of the Pacific Ring of Fire. Little vegetation is found – mainly mosses and lichens. Krill, whales, leopard seals and penguins are the main species found in the Southern Ocean.

35 Temperature change in the Southern Ocean is impacting oceanic ecosystems, for example decline in species and loss of food sources. There is retreat of glaciers and ice sheets in some areas (those fringing the peninsular) and advance in others (to the east of Antarctica). Ocean acidification is disrupting food webs.

36 Tourism is strictly controlled and impacts are monitored. Evidence shows low levels of impact, very little litter, and landing sites with low levels of wear and tear.

37 Summer tourists arrive at peak breeding times, land-based installations are clustered, therefore concentrating their impact, and demand for fresh water is difficult to meet.

38 CO_2 in the atmosphere creates carbonic acid, which makes the ocean less alkaline.

39 Because resources are running out in other locations, some countries (for example Japan and Russia) feel that a low level of careful exploration and usage of resources in Antarctica should be possible.

Chapter 8

1 Location, locale and sense of place.

2 Globalisation is making places more homogenised – more uniform or similar. Some believe that this has been the result of global capitalism, eroding local culture and localised identities.

3 A response to globalisation – it centres on the promotion of local goods and services in an effort to retain local cultures and identities.

4 Through language, dialect, culture and life experiences.

5 Through sociability, access to resources, activities and image – to provide a safe and stable place in which to live.

6 The sense of attachment and belonging that people feel or don't feel for a place. Also, the social, cultural and economic factors – for example, a high wage earner in London will have a completely different perspective from that of a low wage earner who cannot fully engage in the social and economic opportunities and experiences of the city.

7 Exogenous factors are those of external origin – the shifting flows of people, capital and resources. Endogenous factors are those of internal origin – location and physical geography as well as social, economic and demographic factors.

8 Deindustrialisation is the reduction of industrial activity in a region or country. It affects employment and thereby social and economic factors. It can also impact the physical environment – perhaps in terms of derelict sites and disused industrial buildings but also in terms of regeneration and reimaging when areas are redeveloped.

9 From direct experience or from relayed experience from other sources such as media.

10 Any two from: to attract jobs and investment, to host a major event, such as a major sporting event, or to regenerate an area that has become rundown.

11 a) Rebranding is used to give a place a more positive identity.

   b) Re-imaging involves marketing and the promotion of a new identity.

   c) Regeneration is a long-term redevelopment process often adopted in areas of economic decline.

12 Any two from: places that may have suffered from a negative image in the past due to economic decline, crime, violence, the impact of a hazard event, social unrest, deterioration of the built and/or physical environment or a lack of social and/or economic opportunities.

13 Any three from: local government, central government, private entrepreneurs, architects, planners.

Chapter 9

1 A shift in the economy from primary to secondary and tertiary activities; more people living in built-up areas; the physical spread of built-up areas.

2 As most of the growth is in Africa and Asia, it will be a major challenge to keep pace with the rate of growth and provide the resources required to support this growth, for example infrastructure, increased water and electricity supply. It will also be a challenge to maintain sustainable growth with

the necessary planning to ensure that the physical environment is protected.

3 Rural to urban migration has occurred on a large scale, with a high influx of people into urban areas but a lack of jobs and housing to accommodate the rise in numbers. There are natural increases in population numbers as the death rate declines but birth rate remains high. This leads to a youthful population, but again the resources in terms of jobs and education need to be in place.

4 Counterurbanisation refers to the movement of people to small towns and commuter settlements, some of these will be in rural areas. Urbanisation is the increasing proportion of people living in cities.

5 Regeneration is the main process by which people are attracted back into urban areas. It includes housing renewal, improved services, development of old industrial sites and job creation.

6 a) A city with more than 10 million people.
   b) A large city that has global influence in the service sector.
   c) A continuous built-up area (conurbation) of more than 20 million people.

7 Many of the activities associated with globalisation take place in urban areas, for example growth in the secondary and tertiary sectors as a result of increased communications and trade.

8 Processes include: natural population growth and the impacts of a youthful population, where young people are attracted to urban areas for jobs and a perceived better lifestyle; also, rural–urban migration, which is triggered by both push and pull factors, for example pressure on natural resources in rural areas and the attraction of potentially higher wages and education in urban areas if these can be accessed.

9 The 'informal' sector of an economy relates to work that takes place without formal contracts and payment of income tax. It involves a cash economy. It is important to some urban economies as it allows work to take place on a small scale, where there is perhaps not the consistency and scale of employment to warrant formal contracts. However, it allows low-wage earners to gain income and therefore it generates a flow in the economy as spending power is increased. The term has now been extended to include workers whose jobs are not regulated and protected by the government.

10 Causes include: mechanisation and a resulting decline in manufacturing jobs; cheaper labour being available abroad in rapidly industrialising countries, for example India, China, Taiwan; cheaper labour lowers production costs and so makes manufacturing abroad cheaper.

11 Deindustrialisation leads to structural unemployment. If a city is unable to diversify and create new jobs then it will suffer the effects of deindustrialisation for a longer period of time due to the downward multiplier effect – unemployment, low levels of disposable income, low levels of spending, saving and investment – leading to economic decline.

12 Deindustrialisation is a loss of manufacturing jobs whereas decentralisation is the outward movement of people and activities from established city centres.

13 When decentralisation occurs it often leaves behind unemployed and underemployed people in concentrated areas where new jobs have not been created. The negative effects of a lack of economic stimulation lead to multiple deprivations – unemployment, poverty, poor education and health services, high crime levels – so enhancing inequalities in cities.

14 Rental values and property prices are cheaper than central locations. Also, there is more space for facilities, such as car parking for commuters, and easy access from by-passes and ring road developments.

15 Increased affluence, thereby increased demand for leisure and services. Increased training and education leading to more service sector professional jobs and a decline in secondary sector employment.

16 High wage earners contribute to local tax revenues, so services improve; through their spending (as they have more disposable income) the retail and leisure sectors grow. Processes such as the gentrification of properties improve the physical environment, which attracts inward investment and stimulates an upward multiplier effect.

17 Multi-functional, has high-quality education services and high levels of employment in services. It acts as a regional capital and is the location of the headquarters of financial companies and TNCs.

18 A graphical model that seeks to explain land use patterns in urban areas by relating location decisions to the rental value of land. Retail functions keen to locate in the centre of the city – the CBD – will pay high rental value for such land. They will be able to outbid land uses such as manufacturing.

19 Through the impact of urban decision makers such as planners and local and national governments, as well as strategies for re-imaging and regeneration, and economic location incentives such as enterprise zones.

20 The buying and renovation of properties in rundown areas by wealthy individuals. Housing improvement then leads to regeneration of an area.

21 a) An up-market, gated housing development.
   b) A self-contained settlement beyond the city boundary.

22 The wealth gap between rich and poor; it can occur in cities across very small areas.

23 When an area suffers decline, usually triggered by economic factors such as industrial decline, job losses and unemployment. Due to low incomes, local spending and local tax revenues decline, services deteriorate, retail and leisure facilities close due to a lack of local spending power, unemployment rises, and those who are upwardly mobile or who are able to find employment elsewhere move out. The physical environment

Check your understanding and progress at **www.hoddereducation.co.uk/myrevisionnotesdownloads**

suffers from a lack of upkeep and investment, confidence in the area is lacking, no inward investment is attracted and further decline occurs.

24 a) When urban areas have higher temperatures than surrounding areas.

b) When wind velocities increase when air flows through an increasingly narrow gap, for example between two buildings.

c) Another name for reflectivity. A surface with a high albedo effect would be ice, which reflects heat. Darker surfaces absorb heat.

25 Causes include: buildings and road surfaces absorb more heat; there is more cloud cover due to more dust particles and more pollution output; cloud absorbs outgoing radiation and produces a further warming effect; urban areas dispose of surface water quickly, reducing the cooling effect of evaporation; heat comes from Industries, buildings and cars.

26 The emission of pollutants from cars and heavy industry acts as condensation nuclei, leading to the formation of cloud; this leads to more precipitation. Also, because urban areas heat up in summer, the high temperatures can lead to uplift of air and bursts of convectional rainfall.

27 The uneven surface of buildings produces frictional drag. High-rise buildings channel air through the gaps or canyons between them. Upward convection processes can draw in air from cooler surroundings. When air flows between buildings, the air movement is affected by the Venturi effect, where the pressure within the gap causes the wind to pick up speed, forming a gust.

28 When cool air is trapped below warm air. Normally air temperatures cool as distance from the land surface increases, therefore an inversion is the opposite of the normal state. Where an urban area is located in a lowland or dip between hills, cool air can descend into the dip and become trapped by a lid of warmer air above it.

29 Particulate pollution; temperature inversions trapping pollution; photochemical smog which is caused by cars and pollution.

30 Urban growth is rapid and often there are not the resources, such as waste disposal, to keep pace with the development. Waste can therefore often be left lying around without the necessary level of collection and safe disposal services. Waste disposal is often unregulated.

31 We are running out of space for some methods of waste disposal, for example landfill. Therefore, new methods of disposal must be found, or the waste needs to be transported elsewhere for disposal and this increases the cost. Also, waste is being produced in increasing quantities and so the cost of disposal rises as more waste needs to be dealt with.

32 If not treated properly, waste can lead to many problems, including: waste is a large source of methane – a powerful greenhouse gas; waste contributes to water, ground and air pollution; untreated waste can lead to health and respiratory issues and so it is important that it is dealt with safely.

33 a) Land previously used for industry or some commercial purposes.

b) An area of previously undeveloped land.

c) The removal of pollution or chemicals from the ground so that it can be safely developed.

34 The ability to raise living standards and improve quality of life without compromising the needs of future generations.

35 It involves minimising the ecological footprint of urban areas, improving the quality of the environment, reducing deprivation and ensuring a sound economic base.

36 By putting in place some of the following measures: reducing waste (recycling and reuse); reducing the quantity of resources needed; using renewable energy; making building design more energy efficient; regenerating brownfield sites (land that has already been built on); traffic management and reduction.

37 It refers to living conditions. It is a broad term that incorporates many aspects of (urban) living – natural environments, employment opportunities, cultural opportunities and security, for example.

Chapter 10

1 a) Polluted water will spread disease and reduce population growth, impacting the death rate and also vulnerable young children.

b) Damage to wildlife and their habitats will impact natural ecosystems and may affect food supplies, causing a decline in populations where there is a high level of dependency on the animals available within the surrounding area. In some countries, the destruction of wildlife and their habitats may be due to expansion of commercial agriculture which supports population growth through increased food supply.

2 A process whereby nutrient enrichment in water (streams and rivers) leads to a fall in oxygen levels due to the work of aerobic bacteria, and the subsequent death of species that are dependent on oxygen supply.

3 Low rainfall; high rates of evaporation and transpiration.

4 Population distribution is the spread of population across an area, that is the pattern of where people live. Population density is the average number of people living in a specific area, usually expressed as people per square kilometre.

5 Birth rate, death rate and migration.

6 Because 'distribution' of food does not get food supplies to where they are needed and in some areas there is a surplus of food and then waste.

7 Human resources (population numbers), skills, technology and investment in agricultural infrastructure.

8 A measure of the efficiency of agricultural production. It is measured as a ratio of the outputs to inputs in an agricultural system. Investment in technology improves the TFP.

9 Through: growing disease- and drought-resistant crops; improved efficiency; greater use of technology and machinery; the use of high-quality animal feeds.

10 In polar areas, most food production involves a hunting/gathering technique. Indigenous people are able to make use of the sources of food available to them which are animal based, for example reindeer for meat and milk, and fish. In tropical monsoon climates, there is the potential for intensive cultivation of crops, such as rice and wheat.

11 Effects include: increased water stress on crops; yields of rice, wheat and maize in particular would decline; more severe and unpredictable climatic events will lead to a range of impacts, such as flooding and drought, and events such as cyclones will destroy crops; warmer temperatures may extend growing seasons in more northerly latitudes.

12 a) When small channels of water called rills combine to form deeper 'gullies'. Water is channelled through these gullies and, as it flows, it erodes the soil.

   b) When overland flow transports topsoil in a uniform way.

13 Very small particles of material that have been lifted in the wind and carried away from the erosion site, settling elsewhere.

14 Most organic nutrients are stored in the vegetation itself. Due to the warm, wet climate and rapid growth rates of vegetation, there is rapid recycling of nutrients and a very low store of organic matter in the soil, making the tropical rainforests difficult to farm productively without careful management.

15 Salinisation is an increase in the amount of salts in the soil. They are brought to the surface due to high rates of evaporation and transpiration, which combine with low rates of precipitation and poor soil drainage. If irrigation systems are poorly managed, they can supply more water than the crops use and this exacerbates the process of salinisation.

16 Food security, as defined by the UN Food and Agriculture Organization (FAO), exists when 'all people at all times have physical and economic access to sufficient, safe and nutritious food that meets their dietary needs and food preferences for an active and healthy life'. It is a complex term with many dimensions which cover more than just food production.

17 Availability, access, utilisation and stability of food supply over time.

18 By promoting healthy eating and by education on food nutrition.

19 Mortality relates to death. Morbidity relates to illness and disease.

20 A model suggesting that over time as a country develops there will be a transition from infectious disease to chronic and degenerative disease as the main cause of death.

21 Education can impact in several ways: people can become better educated about disease prevalence and how to avoid or treat diseases; better education can also have economic impacts in terms of the jobs people do and the lifestyle and quality of life they have. All of this can improve their well-being.

22 To update the model with possible future trends such as: potentially entering a stage of 'an age of healthy living' due to good diets and exercise or 'an age of obesity and inactivity' as a result of modern developments in technology and growing affluence – leading to higher levels of car ownership. Another suggestion for the fifth stage is the possible emergence of new infectious diseases.

23 Links between climate and disease include: drought – leading to the potential for famine as a result of poor harvests; flooding – leading to water-borne diseases and respiratory illness; seasonal adjustment disorder – resulting from the lack of sunlight in winter months in parts of northern Europe; malaria – spread by mosquitoes that require temperatures of 16–32°C.

24 Within the home, there may be a rise in respiratory infections where households are reliant on fuelwood for cooking and heating purposes. Women and children are particularly at risk.

25 Vital rates for the measurement of population and include birth rate, death rate, population growth rate, total fertility rate, net replacement rate and infant mortality rate.

26 Births, deaths and migration. Population change of a city or region is a result of births minus deaths plus or minus migration.

27 a) The average number of children borne per woman – depending on survival rates it is most likely to increase a population.

   b) The number of children each woman needs to have to maintain current levels of population. As the measure relates to maintaining current levels of population, the impact on population growth should be minimal.

   c) The number of children who die before their first birthday per 1000 live births per year. A high infant mortality rate reduces population growth if 'replacement' children have low survival levels.

28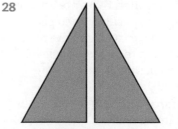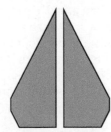

Youthful population   Ageing population

29 Youthful population: an increased dependency of the very young on a smaller proportion of the working population and more spending on healthcare and education for the young. As the young people progress to working age (16–64), there will need to be investment in training and a thriving economy to provide work, otherwise there will be high levels of unemployment with further economic impacts.

Check your understanding and progress at **www.hoddereducation.co.uk/myrevisionnotesdownloads**

Ageing population: increased dependency on a smaller proportion of economically active adults and a large proportion of elderly. This may lead to a lower GDP and less economic growth and a large proportion of spending on healthcare and pensions.

30 The benefit a country experiences when its working population is greater than the dependent age groups – the very old and the very young. The result is a boost in the economic productivity.

31 To ensure that there are sufficient jobs available as working people contribute towards tax revenues and generate economic growth through their spending power. If there are not enough jobs available, the state has to pay out more financial support to the unemployed.

32 A refugee is a person fleeing, perhaps due to war, persecution or civil unrest. An asylum seeker is a person who has fled their country of origin and is seeking asylum under the 1951 Convention. People fleeing an area are initially refugees, they seek asylum in a chosen new country, and only when that application for asylum is granted can they be given legal refugee status.

33 Optimum population exists when there is a perfect/ideal balance between population and resources.

34 Carrying capacity is the maximum population size that an area or environment can sustain indefinitely. Ecological footprint is a measure (usually in global hectares) of the land and water needed to produce the resources that humans utilise and assimilate the waste that they generate.

35 The pollution of water or the over-use of land resources for farming both lead to lower levels of food production, less food for people, famine, and population decline.

36 a) Desertification leads to a decline in food production and potentially increased mortality rates and low population growth.

b) Pandemics on a large enough scale reduce population growth as they increase mortality.

c) Population growth slows with increased affluence because of the high cost of raising children and because women have fewer children due to career aspirations as a result of high levels of education.

37 Fertility rates (an expected global downward trend) and life expectancy (increasing life expectancy globally due to medical advances and improved standards of living).

38 Pressure on the provision of health and social care for the elderly; increased home (domiciliary) care requirements; isolation and loneliness as more older people are living alone, pressure on pension provision; pressure on the working population to build up tax revenues needed to pay for additional health and social care.

Chapter 11

1 The safe, sustainable and reliable flow of resources to maintain existing levels of development and allow future generations to advance.

2 Stock resources are non-renewable and flow resources are renewable.

3 a) Water, soil, forests, wind, tides, waves.

b) Oil, gas, coal, uranium, metallic ores.

c) Metallic ores and water.

d) Fossil fuels.

4 Both have a role to play in advancing development in countries and indeed globally. Human resources provide labour and skills, and capital supports the development of resources for economic advancement and to meet human needs.

5 A resource is any part of the natural environment that can be used to meet human needs. A reserve is that part of a resource that is economically, legally and technically viable to extract.

6 Possible reserves are where there is insufficient information on quantity and quality of the reserve. Proven reserves are economically viable to extract and have undergone a preliminary feasibility study. Probable reserves have a reliable estimate on quantity, grade and quality of resource.

7 An indicated reserve is that part of a mineral resource for which quantity, grade or quality, densities and physical characteristics can be estimated. An inferred resource is when there are assumptions based on some geological evidence of quantity, grade and quality.

8 As low-income countries develop, their demand for resources will rise. Environmental impact of current resource use is a concern; this will increase particularly if there is a lack of regulation. Therefore, there needs to be careful management of resource use so that future generations have sufficient supplies.

9 An area where resources are brought into use for the first time.

10 Sustainable resource development is important so as to not limit the possibilities of advances and development for future generations.

11 They are an important part of sustainable resource development; they balance economic gain against environmental impact and have a role to play in regulating resource development.

12 Due to advances in technology allowing more efficient and sustainable use of resources.

13 Asia is where economic growth is rising at the fastest rate and there is an emerging affluence of the populations in countries in Asia which will drive demand and consumerism.

14 Three from: increased economic growth; increased affluence and rising levels of disposable income in a country's population; higher levels of domestic car ownership; the growth of transport infrastructure to support rising levels of production.

271

My Revision Notes: AQA A-level Geography Second Edition

15 Energy security is the ability of a country to have uninterrupted energy at an affordable price. It is harder to achieve in some countries than others. For example, a country like Japan that is economically developed but has few indigenous resources of its own will consume more energy than it can produce. Other countries may be resource rich and have a surplus, for example oil-rich countries in the Middle East.

16 Refers to a country which has an abundance of natural resources (for example minerals) but which experiences minimal economic growth due to the poor management of these resources. This can be due to:
+ an over-dependence on the production and export of a resource that brings low financial return
+ a lack of investment in other sectors, for example manufacturing
+ over-dependency on TNCs.

17 They can bring about a reduction in production costs and increase supply; if demand falls, this will result in a lower price.

18 They have a water surplus due to efficient management of water resources; such countries include the USA and Russia.

19 Physical water stress is a lack of water to meet demand. Economic water stress is the lack of means to make use of the water that is available, that is it cannot be extracted, transported or treated.

20 If evapotranspiration rates are high, then rainfall will evaporate before it can be captured and made use of. This happens in countries with a hot climate where there is rainfall but it evaporates or where it is lost through transpiration.

21 The drainage basin system receives inputs of precipitation and outputs are channel flow, runoff, evapotranspiration and groundwater flow. Local climate and geology affect these inputs and outputs. The relationship between inputs and outputs determines water supply.

22 Advantages include: they are an effective means of storage; they are non-polluting; they are a method of flood control; they can be used as a source of energy; they can generate further revenue from tourism.

Disadvantages include: the loss of land that has to be flooded to create the reservoir – this includes farmland and settlements; flooding land removes natural habitats; disruption of the hydrological cycle.

23 This involves the removal of salt from sea water. The two main methods are reverse osmosis (filtering of seawater at high pressure) and distillation (water is boiled and the salt is left behind).

24 a) A measurement of the water required in the production of agricultural and industrial products. If a country imports a food resource, such as lettuce, which requires a large input of water, then it has a virtual water value.

b) Water that has been used for cleaning and washing but that does not need sewage treatment so that it can be reused for irrigation, for example.

25 Rivers and drainage basins cross international boundaries and the effect of one country's actions can impact water supply in a neighbouring country, leading to disputes. A number of water conflicts exist globally, for example Tigris–Euphrates – conflict between Syria, Iraq and Turkey over Turkey's water management project which diverts water supplies; the River Nile – conflict between Ethiopia, Sudan and Egypt over control of the Nile's headwaters.

26 Depth, thickness and quality of coal.

27 Many developed countries cannot satisfy their oil requirements and have to import large quantities of oil, which leaves them in a vulnerable political position as supplier countries can threaten to cut supply.

28 There is an increase in proven reserves of shale gas and a lot of money has been invested in the transportation and distribution of gas supplies.

29 Advantages: it is efficient, has large reserves and is less polluting than fossil fuels. Disadvantages: it requires high levels of investment and technology and careful handling of waste; it is difficult to find suitable sites to build them and nuclear power stations are a potential target for terrorism.

30 When a country retains an existing energy mix due to the economic (cost) or technical difficulties of changing.

31 Diversification spreads the risk should there be changes in supply, rising costs of production or geopolitical tensions surrounding a certain energy source.

32 TNCs, such as BP and ExxonMobil, influence oil supplies through activities in exploration, distribution and production.

33 Because of the diversified nature of copper's end use in electrical products, construction and vehicles. All three are important indicators of economic development and well-being.

34 The main environmental impacts of ore extraction are: loss of natural habitats, noise and dust pollution, and toxic leachates.

35 Economic gains may outweigh environmental destruction, which will allow further impacts on the natural environment. Emerging economies which are resource-rich, for example in Latin America, may not have strict environmental protection legislation in their desire to progress and utilise resources. Specific concerns relate to resources in Antarctica that are protected.

36 Two of: osmotic distillation of seawater; saltwater greenhouse technology; small-scale appropriate technologies, such as water cones to distil seawater in small quantities.

37 Two of: the influence of TNCs potentially overriding government decisions; the increasing reliance of developing countries on TNCs to develop their energy resources; increasing pressure on reserves in the Arctic and Antarctic, and pressure on the concept of the global commons.

Check your understanding and progress at www.hoddereducation.co.uk/myrevisionnotesdownloads